Philosophy of Engineering and Artifact in the Digital Age

Philosophy of Engineering and Artifact in the Digital Age

Edited by

Viorel Guliciuc and Emilia Guliciuc

CAMBRIDGE SCHOLARS
PUBLISHING

Philosophy of Engineering and Artifact in the Digital Age,
Edited by Viorel Guliciuc and Emilia Guliciuc

This book first published 2010

Cambridge Scholars Publishing

12 Back Chapman Street, Newcastle upon Tyne, NE6 2XX, UK

British Library Cataloguing in Publication Data
A catalogue record for this book is available from the British Library

ISBN (10): 1-4438-1970-0, ISBN (13): 978-1-4438-1970-1

TABLE OF CONTENTS

Letter Addressed by Luciano Floridi to the PHEADE 2009 Participants

University of Hertfordshire

l.floridi@herts.ac.uk
www.philosophyofinformation.net
Tel. +44 (0)1865 271970
Fax: +44 (0)1707 285611
De Havilland Campus
Hatfield, Hertfordshire,
AL10 9AB, UK

10 December 2008

To the Organisers
Suceava's International Workshop
PHEADE 2009 – Philosophy of Engineering and Artifact in the Digital Era /2009

Dear Colleagues,

it is with great pleasure that I send you this message of greetings and congratulations on behalf of the International Association for Computing and Philosophy (IACAP, www.ia-cap.org).

Following Prof. Peter Boltuc's welcome suggestion, I have been following initiative in support of PHEADE 2009 and publicised within our association, as one deserving careful attention from our membership. It is a splendid and very timely idea.

We know that the world is changing dramatically and at an unprecedented speed. Never before has humanity experienced so many and such radical changes and challenges, in so many contexts, in such a short time. The transformations of our time so often depend on new technologies, which are themselves constantly evolving. It is a bubbling new reality in need of understanding and intervention. So it seems clear that our future requires a new alliance between philosophy and engineering, perhaps more in line with Francis Bacon than with Plato.

What unites us, philosophers and engineers, is our interest in understanding and building systems, whether purely conceptual or physical. We, as builders, carefully and painstakingly construct the conceptual and practical frames and tools that enable humanity to make sense of the world and to flourish in it.

It is for this very reason that IACAP has always supported a robust and fruitful dialogue between philosophy, engineering sciences, and technology. It is hard to think of a more crucial area of philosophical investigation in our society, which we define, after all, as the information society.

All this explains the very high appreciation and esteem that I would like to express, on behalf of IACAP, towards the organisers and the staff of PHEADE 2009. It is an impressive initiative, which we all hope may have a bright future.

To conclude this short message, allow me to wish to all participants a fruitful and pleasant meeting. I am sure you will enjoy the talks and take full advantage of the many opportunities for new interactions and learning offered by such a stimulating environment.

All my best regards,

Luciano Floridi

Professor Luciano Floridi
(Laurea Rome "La Sapienza", MPhil and PhD Warwick, MA Oxford, Gauss Professor, Akademie der Wissenschaften, Göttingen)

Research Chair in Philosophy of Information
Director, Research Group in Philosophy of Information
University of Hertfordshire

Fellow of St Cross College
Director, Information Ethics Research Group
University of Oxford

CHAPTER ONE

EXPLORATIONS IN THE PHILOSOPHY OF ENGINEERING AND ARTEFACT

PHILOSOPHY, ENGINEERING AND TECHNOETHICS

VIOREL GULICIUC

1 Engineering Crisis and Technoethics

Most academic subjects have a meta-theory, a philosophy, as the case with the philosophy of science, the philosophy of mathematics, the philosophy of logic, the philosophy of law, the philosophy of communication, the philosophy of education, the philosophy of history and so on.

In the second half of the 20th century, in times of crisis, both for philosophy and for engineering, humankind rediscovered how the highest *techné* had a definite theoretical dimension and the highest *theoría* is also a practice. Expressions such as "practical theory" or "theoretical engineering" became common. However, this was possible only after the major crisis of philosophy itself, expressed in postmodernism, and the major crisis of engineering itself, when considering the so called emergent technologies (and especially robotic, nanotechnology and genetic engineering).

The workshops organized worldwide on the philosophy of engineering[1], *seem to suggest that the interest is not for the philosophy* in *engineering, but for the philosophy* of *engineering.*

> "Engineers are ... the unacknowledged philosophers of the postmodern world"

writes Carl Mitcham, (1998, 27).

The *ethic concerns* on the use of the emergent technologies (Joy, 2000) are waking up the interest of the philosophers too; the logic and the

---

[1] After the series ***Philosophy of Engineering*** workshops organized by the Royal Academy of Engineering and ***WEP 2007*** (Delft) and ***WEP 2008*** (London), we could mention ***Working Towards a Philosophy of Engineering,*** a seminar organized on 25 January 2006, Loughborough University or the ***PHEADE 2009 (Philosophy of Engineering and Artefact in the Digital Age - 2009)*** workshop organized at "Stefan cel Mare" University (Romania).

philosophical methodology are more and more appealing for the engineers etc.

Day after day, *theoría* and *techné* meet in various ways, because, at its origins, the highest *theoría* has the *techné* inside and the *techné* itself is a *theoretically* designed activity.

We could say, based on the advances in modern technology, that engineering has provoked dramatic changes in contemporary society; that we are living in a society of engineering. So, engineering could be of bold ethical, contextual, and cultural interest.

Let us remember the brutal loom of ethics in engineering when discussing the *Q-ring* problem, after the crash of the space shuttle Challenger, on January 28 1986. Seven lives was the price for responsibility, unjustly paid by the engineer that designed those gaskets, forever after having a life of conferences on the ethics of engineering.

Nowadays engineering is enriched by the ethical debates from IT, nanotechnology, robotics, genetic engineering, bionic systems and communication technologies. In fact, one could observe that the number of the events organized under the "umbrella" theme of ethics is almost asymptotically increasing. There is interest for topics related to the relationship between engineering and theology.

These situations seem to express some ground fears and awareness of the human condition and future in a technologic society…

Even its roots cannot be separated from the philosophy of technology (Vermaas, 2007, 25), it is only under globalization, beginning with the transmodern era, philosophy of engineering becomes an emergent domain (so, *after* the postmodern crisis of philosophy[2] and the first moments of the *crisis* of engineering itself (Goldberg, 2007)).

---

[2] Let's remember that in the very *Introduction to The Postmodern Condition: A Report on Knowledge* Jean-François Lyotard writes: "Simplifying to the extreme, I define postmodern as incredulity toward metanarratives. This incredulity is undoubtedly a product of progress in the sciences: but that progress in turn presupposes it. To the obsolescence of the meta-narrative apparatus of legitimation corresponds, most notably, the crisis of metaphysical philosophy and of the university institution which in the past relied on it. The narrative function is losing its functors, its great hero, its great dangers, its great voyages, its great goal. It is being dispersed in clouds of narrative language elements--narrative, but also denotative, prescriptive, descriptive, and so on [...] Where, after the metanarratives, can legitimacy reside?" At: http://www.idehist.uu.se/distans/ilmh/pm/lyotard-introd.htm

> Here I argue that a key reason we are now meeting in Delft, at least from the perspective of engineers and engineering educators among us, is that, like physics at the turn of the 20th century, engineering is in considerable crisis because the engineering paradigm of WW2 and the cold war is unable to effectively design artefacts of a postmodern creative age. (Goldberg, 2007, 36).

Despite the fact that it is so new, there are already some newer branches of the philosophy of engineering itself, *largo sensu,* for example, the *applied philosophy of engineering*, or *semiotics of engineering*, or several distinct possible philosophical perspectives (Mitcham & Mackay, 2007).

Based on its ethical background, the philosophy of engineering is a transmoderne discipline. So, it is not by chance that precisely "in the age of spiritual machines", when the questions of technoethics are more and more important in our lives, the interest for "a relatively uncharted topic – the philosophy of engineering" is in a rapid *éclat* and, consequently, "the philosophy of engineering is a discipline that is just beginning to emerge" (McCarthy, 2008, 48).

Attempts to explain this (commonly) unexpected association between philosophy and engineering were made starting from the association between philosophers and engineers, as it has already been made occasionally: an engineer turned into a philosopher (Wittgenstein), or a philosopher turned into an engineer (Mark Bedau). However "a formal meeting of engineering and philosophical minds is a rarity" and raises the question "why it has happened at this moment in history"? (Goldberg, 2007, 35). However, let's agree that the theme of the incorporation of engineers in philosophy and philosophers in engineering, even though it is often pointed out, is hardly exploitable.

The main questions when a new field of research is opening are focused on what is distinguishing it from other research domains. From a traditional, common perspective that means to analyze the objects *(largo sensu)* related to that research, the knowledge characterizing it, the specificity of its methods and logic, and the ethical implications of the studied domains.

In the effort (epistemographic) of positioning the philosophy of engineering, among other research fields as philosophy of technology, philosophy of science or philosophy of creation, some of the main questions we will have to consider are:

What are the fundamental concepts in engineering that need to be defined in order to underpin a philosophy of engineering?
What concepts from philosophy should be used to help build a philosophy of engineering?[3]

Of course, one of the best strategies would include a synopsis of all or majority of the questions on and of the philosophy of engineering. However, even in an emergent field of study, the velocity of publishing papers definitely surpasses the time needed to read and analyse their contents (and their questions). Obviously, this paradoxical situation will not stop the essay here.

Our *thesis* is that *engineering is an activity profoundly related to philosophy, by its specific type of use of the universality and by its ethical (not simply moral) dimensions.*

2 Etymology Games

Some of the contemporary researches on *philosophy of engineering* seem to focus the discussion on (so, to reduce it to) the original relationship between *episteme* and *techné, knowledge* and *craft*. The Greek word *techné* has to be philosophically distinguished from the word *episteme* (Parry).

The question is: how could we use the lexical dimensions of philosophy and engineering? We will follow Aristotle's recommendation for the knowledge of the words' meanings. That means we will fully play with the words and their meanings, because the sense is always in the play.

The term *philosophy* is inherited from the ancient Greeks. It was gradually established, because philosophy itself gradually became distinct as a form of spiritual life (Eduard Zeller). Its uses are multiple: "1. *love of knowledge and wisdom, pursuit thereof, speculation, study*"; "2. *the systematic treatment* of a subject, *investigation*"; "3. *philosophy, the investigation* of truth and nature".

Philosophical knowledge was classified in *theoretic*, *epistemic* or *systematic*. As *theoria*, philosophy was less knowledge, in the contemporary use, than action of contemplation. The Romanian logician and historian of logic Anton Dumitriu, has proposed a distinction between the *inesse philosophy* and the *dicitur philosophy*, as based on *acategorical*, like intuitive, continuum knowledge or based on *categorical*, rational, discriminatory, common knowledge (Dumitriu, 1977).

So, philosophy was an *activity (= theoria)* and not a result/description.

---

[3] See: http://www.engsc.ac.uk/nef/events/philosophy0106.asp

Moreover, "love of wisdom", "love of knowledge", "investigation of truth" does not *necessarily* imply philosophy is searching for the truth, nor that it is a knowledge. From this perspective, those who claim that science is searching for the truth, that science is (*stricto sensu*) knowledge, may have some right. This is why the philosophical "truth" and "knowledge" are quite different from the scientific ones.

Engineering was defined (by the American Engineers' Council for Professional Development, in 1941) as:

> "the creative application of scientific principles to design or develop structures, machines, apparatus, or manufacturing processes, or works utilizing them singly or in combination; or to construct or operate the same with full cognizance of their design; or to forecast their behaviour under specific operating conditions; all as respects an intended function, economics of operation and safety to life and property".

De facto, *engineering* is derived from *engineer* (Dear, 2001, 25) one who operates an *engine*, and was "a constructor of military engines." The word *engine* itself is old enough as it is derived from the Latin word *ingenium* (that is why, maybe, engineers naturally show *genium,* but also *ingenuity*) which is deeply related to the Greek word *techné*. So, engineering itself is rooted in *techné*. That is why, exploring the etymologies of *techné*, we could discover some original characteristics of engineering itself. *techné*

The ancient uses of *techné* are:

1. *Art, skill, craft in work, cunning of hand* (of metal-working; of a shipwright; of a soothsayer).
2. *The way, manner* or *means whereby a thing is gained*, without any sense of art or craft.
3. *An art, craft, trade*, to know his *craft*; *to practise* it; to learn a thing *professionally.*
4. *An art*, i. e. *a system* or *method of making* or *doing*.[4]

As a method, *techné* (engineering) seems to be non-epistemic (because *epistéme* is not *organon*), meanwhile as knowledge (theory) it always keeps inside the teleological dimension of the making, because it is more systematic, than epistemic or theoretic.

[4] Search on *Perseus* at: http://www.perseus.tufts.edu/

However, "given that many scientific theories that seek knowledge about the world involve engineering in that endeavour, engineering should surely be of interest to philosophers", there are pertinent questions to be asked about: "what exactly is the role of manufactured objects in finding knowledge?", "How reliable are they?" observes Natasha McCarthy.

Engineering is delivering knowledge by a much more direct route than by aiding science. And *is obscured by the single word 'knowledge.* When we take it into consideration it is clear that engineers seek to acquire knowledge in all of their endeavours.

So, "engineering is 'know-how'". As a consequence engineering yields

> highly successful knowledge about how to control materials and processes to bring about desired results. It is a way of getting to the nature of things – a voyage of discovery as much as science is. Hence engineering provides a useful case study for philosophers inquiring about the status of human knowledge. (McCarthy, 2006, 49)

And yet:

> Although there are revolutions in engineering, the products of engineering knowledge are not going to be overturned in the way that some scientific theories have been." This is why "that knowledge of what works, the 'know-how' that engineering provides, is secure knowledge. Engineering knowledge is also genuinely cumulative – improved all the time by building on, and not re-writing, what went before. (McCarthy, 2006, 49)

Yet, even if engineering is securing knowledge, it is not also a search for the truth, but for better and better artefacts: *the artefacts produced by engineers are not "true", but well done.*

Resuming, *when philosophy is interested in the action of searching the senses of our knowledge, engineering is interested in the action of securing it.*

For both of them, at the beginning, it was the *action.*

3 What Engineering *is*

For a philosophy of the engineering to formulate a response to this question seems to be a must.

The contemporary debates on the specificity of engineering are privileging two of the possibilities: *engineering is a distinct form of activity (practice)* and/or *engineering is a form of knowledge.*

Considering the first choice, let's agree that there are many types of human activities: thinking, building, writing poems, modelling, singing, designing, crafting etc. Focusing on engineering as an activity, the first question will be: *what type of activity is engineering?*

The answers are various, from *engineering is design*, to *engineering is making*, *engineering is logic of production*, *engineering is "construct[ing] a Bill of Materials for any complex artefact*" [McCarthy, 2006, 50], *engineering is establishing a technological chain of operations* or *engineering is securing the knowledge* necessary to produce an artefact.

The attempts to reduce engineering to one activity, confesses a strong belief in the necessity of reducing any complex human activity to a simple, unique activity. In fact, this belief is a presupposition: we are *presupposing* engineering is *an* activity.

But what if engineering is a group of activities?

Indeed, an engineer is simultaneously doing several activities: multitasking thinking, imagining, designing, planning the logistics, making etc.

Our temptation of reducing engineering activities to one single activity illustrates the belief that complexity could be (so, has to be) reduced to simplicity. Or, sometimes, the *complexity is irreducible*. Let's observe that we would not be able to imagine, design and produce complex artefacts /systems of artefacts, if we would not have, from the very beginning, the capacity to understand and manage complexity itself.

So various are the engineering activities, that it is a sign of complexity, and some authors are speaking not about engineering, but about engineering(s) and consequently not about philosophy of engineering, but about philosophies of engineering(s).

That is, we could more correctly define engineering as the complex of activities related to the artefact.

Because *artefact* is a central concept in engineering, so it is in the philosophy of engineering, too. Nowadays, the word is used with the following senses: "any object made or modified by a human being", "a kind of tangible result produced during the development of a software", "a human-made object integrating information about the culture of its creator and users", "a product of human being(s) in a social behaviour", "an object from the digital environment".

The concept of *artefact* is valued as important by Luciano Floridi (2005) (when discussing on the constructionism and knowledge in engineering), by Ibo van de Poel (when presenting the *Delft University's researchers* conceptions of the two descriptions, physical and intentional, of the artefact[5]), by Natasha McCarthy (2006) (when considering that artefacts are aims of engineering and analyzing the philosophical dimensions of the "inherent complexity of engineering artefacts") etc.

On the other hand, engineering is also considered as a specific form of design.

> The subject of 'Engineering' can be divided into two activities of analysis and synthesis. Analysis (& research) is really Science or the study of materials, actions, life etc. to better understand our world. We can call it 'Engineering Science' if the study focuses on materials, processes and material actions. However, when we start taking this knowledge and applying it to improve the quality of life we are synthesizing knowledge. We are now being creative with our knowledge. This is *design* and is *fundamental* to *engineering. Engineering* is *design.* Research & analysis is Science. Both Engineering and Science are important; but knowledge alone is of no consequence to the future of life if it does not manifest itself into material significance through *design.* (Green).

Considering the second choice, there also are various forms of knowledge. The *relationship with knowledge distinguishes philosophy, science and engineering.* Engineering is commonly considered as a specific form of knowledge. However, it cannot be reduced to either scientific knowledge or to the empirical one: if in philosophy we are in the pursuit of questions like: *Have we a genuine knowledge about the world?, How are we acquiring that knowledge?, Can we ever be fully confident it is genuine?*; in science, the aim is to find truth and to acquire genuine knowledge (McCarthy, 2006), and to observe that engineering is an important part in acquiring our knowledge, because theoretical science's search for knowledge, "supported by experimental data", would not be attainable without it. So, engineering is not "in the truth business", as we are considering science is. It is not a form of scientific knowledge, but a form of securing (scientific) knowledge.

It seems that a reductionist mechanism prevails in the existing responses on what is engineering, when reduced to a distinct form of practice (activity), or to a form of knowledge.

---

[5] During the seminar organized on March 27, in 2007, on *Philosophy of Engineering* at the Royal Academy of Engineering.

This situation is multi-philosophically highly significant. Let's explain. Mircea Eliade observed that the Human Being has several types of insertion into reality, into being. A modern researcher would say that our consciousness is already structured when being-toward.

As scientists, we are searching for the truth; as artists, we are sensible to the beauty; as religious men, we are looking for the sacred; as philosophers, we are hunting the sense; as moral people, we are hoping for the right. They could all be considered as *interfaces* with the being.

Because engineering could not be reduced to science, neither to art, or any other human activity, we will not speak about the philosophy of engineering as about the philosophy of art, of science etc

Maybe it is time to consider engineering as the sixth human mode of insertion into reality.

4 Speculations

Reducing philosophy to an exclusive theoretical activity and engineering to an exclusive practical one is not respecting the original meanings of those two types of human activity.

Or, in the original use of a word, we always will find its future uses, because "any provenance is already a future" - "ein Herkunft aber bleibt stehts Zukunft", as observes Jean Beaufret when bolding common themes and ideas of the philosophies of Husserl and Heidegger.).

Let us remember that, in the very introduction of the first book of his famous work ***The lives and Opinions of Eminent Philosophers***, Diogenes Laërtius, notes that "if we may believe the Egyptians, Hephaestus was the son of the Nile, and with him philosophy began, priests and prophets being its chief exponents. Hephaestus lived 48,863 years before Alexander of Macedon ....." (*Laërtius*) [6].

Ignoring for good reasons "if we may believe", from the above quotation (it is not suggested by the Greek text), *Diogenes Laërtius's statement is bolding the common roots of philosophy and engineering.*

---

[6] However, the Greek text, accessible at: http://www.mikrosapoplous.gr/dl/dl01.html#thales would be preferable, for multiple reasons. For example, in the online text of R. D. Hicks translation (1925, Harvard University Press), Hephaestus is replaced by Vulcan. Or Hephaestus was only *grosso modo* the Greek equivalent of Roman god Vulcan and the Egyptian god Ptah (from Memphis). He was the god of *technology, craft, metallurgy, fire, volcanoes a.o.* That is why, the use, in the online version, of the term "Vulcan" instead of "Hephaestus" confuses, because primarily, the Roman god was the god of fire. Their different primary meanings create different philosophical uses.

Those common roots could be found in some of their ancient etymologies. For illustration, let's remember that David, the Armenian, in his ***Introduction to Philosophy***, was reviewing six different definitions of philosophy. Two are particularly interesting for us: one consideration, with Aristotle, is that philosophy is the *study of the being qua being*; other, considering it as *purification for death*. It is about *theoria*, the non-demonstrative, the direct knowledge, and not about the *episteme*, the demonstrative knowledge, the mediated knowledge here.

So, when *philosophy was the wisdom of the study of being qua being, techné* was not only making, but *wise* making, as activity of the master craftsman, as the same Aristotle is bolding in ***Metaphysics***:

> [...] what is called Wisdom is concerned with the primary causes and principles, so that, as has been already stated, the man of experience is held to be wiser than the mere possessors of any power of sensation, the artist than the man of experience, the master craftsman than the artisan; and the speculative sciences to be more learned than the productive (1.981b)

Maybe a research of the specificity of the philosophy of engineering should focus on the dynamic perspective and not on the static one: *what philosophy and engineering are doing* and lesser *what they are / are considered to be.*

This is why *any research in the emerging philosophy of engineering should not forget that both philosophy and engineering are deeply rooted in their common nature of human activities.*

5 Dimensions in the Philosophy of Engineering

As William Grimson observed, there is a huge need now for a systematic approach to the philosophy of engineering. In the mean time, let's also observe that a *study in the philosophy of engineering is not like an epistemographic one* (Dear, 2001)[7], *but like an epistemological one.*

We need, *an inventory of the integration of the philosophy in engineering subjects, into the themes of the philosophy of engineering*, but without limiting their number "through the 'lens' of the five classical branches of philosophy which are taken here to be epistemology, logic,

[7] *Epistemography* as "an enterprise centrally concerned with developing an empirical understanding of scientific knowledge, in contrast to *epistemology*, which is a prescriptive study of how knowledge can or should be made." (Dear, 2001)

metaphysics, ethics and aesthetics", that could be considered, *dimensions in the philosophy of engineering.*

Philosophy of engineering has a metaphysic (ontological) dimension, because it is investigating the nature of the artefact as being, the mereology of artefact components etc. and concepts like *part/whole, structure, function, life-cycle, emergence, process and product, needs and requirements, success and failure, design and planning.*

Here we could formulate the question of whether "there is a common sets of ontological presuppositions and assumptions for all the diversity of activities we call 'engineering'" (Chang). We could agree that among the concepts and objects fundamental to engineering one could find: design, emergence, part/whole, structure, function, process & product, life-cycle, needs and requirements, success and failure etc. essential in managing complexity, as Giancarlo Guizzardi (2005) is founding in the works of Peter Simons.

Philosophy of engineering has a gnoseological (epistemological) dimension because, even it is not a truth business, it is concerned with the nature and scope of knowledge in engineering, and whether knowledge is possible in engineering.

In the search for an answer on the nature of engineering knowledge, a series of questions have to be asked:

> What are the intellectual foundations of engineering? What is engineering knowledge, and what is it to have engineering knowledge? What can philosophers learn from engineering about knowledge? Do engineering and science share a common goal in the quest for knowledge, and do they make equal contributions to our knowledge of the natural world? (Floridi, 2005)

For Luciano Floridi, *constructionism* is the form of knowledge in engineering:

> the constructionist approach of knowledge is based on the presupposition that knowledge itself could and should be reduced to the actions of an epistemic agent. (2005)

Ultimately, the specific condition of human knowledge in engineering is sending us toward the question: *do we have to redefine epistemology and gnoseology or do we change them?*

Philosophy of engineering has an axiological dimension, because the artefacts designing, producing and using are value based activities.

The need for a coherent debate in the rational understanding of engineering design already has a philosophical dimension. From this perspective, engineering is primarily a social rather that a technological discipline. It is informing and influencing societal attitudes and values.

In the *Transcript of Discussions* during the seminar philosophy of engineering (27 March 2007), Ibo van de Poel is speaking about the "value sensitive design". He states:

> this is the idea of building values into the design of products, to make them better, from a moral point of view. It is not only a question of avoiding what is bad with technology but also to do good with technology.

Philosophy of engineering has an ethic dimension, because it is concerned with questions of meta-ethics and ethics applied to all the very different activities we are calling *engineering*.

One could observe that the ethic problems are omnipresent nowadays in the events under the philosophy of engineering, as, for example: "How many people could ecologically live on our planet?" "What are the philosophical implications of this approach?"[8]

When the problems of ethics arise under the larger frame of the philosophy of technology, then the ethic issues debated are multiplying in an explosive way and concepts like risk, uncertainty are more and more important. Moreover, philosophy of engineering is a duty, like knowledge itself is a duty in a technological world (Lorenzo Magnani, 2007).

Philosophy of engineering has an aesthetic dimension, because some of the subject of industrial design deals with beauty, art, enjoyment, emotions, etc.

There is a whole discussion about the need of aesthetic (as it is with logic or ethics) education in the culture of engineering, or the presence of aesthetic values in engineers' artefacts.

> "Engineering is the display of culture. In the engineering practice processes, different times, different places, different ethnic groups, different knowledge and skills background and different aesthetics will all influence the design, construction of the specific projects and even the maintenance of the project results." (Bao, 2007, 76).

Philosophy of engineering has a philosophy of mind dimension, because it deals with the problems of the engineering as making vs.

---

[8] The best example is the book *Abstracts of the Workshop* Philosophy & Engineering 2007, Delft University of Technology, October 29-31, 2007, http://www.illigal.uiuc.edu/web/wpe/files/2007/10/wpe2007abstracts.pdf

engineering as design nature, an extension of the problems of the mind and its relationship to the body.

During the philosophy of engineering seminar of the Royal Academy of Engineering, on the AI and IT, on 11th July 2007, the questions debated were:

> *What can engineering tell us about consciousness? Who decides when we have built a conscious machine - the philosopher or the engineer? What can philosophers and engineers contribute together to the study of, and research into, AI? What issues does the Web, especially the Semantic Web, create for philosophy?*

Philosophy of engineering even has a dimension related to the philosophy of language, because engineers are using language in a particular way.

Maybe that is why Carl Mitcham and Robert Mackay are interested in a linguistic approach as one of the possible sub-fields of research in the philosophy of engineering (Mitcham & Mackay, 2007, 29-30).

Moreover, the use of language could be considered as a technological subject, when accepting that a word has many similarities with an artefact (Monk, 2005).

From such a perspective, it is the time to fully *accept that a word is an artefact.*

Philosophy of engineering has a political philosophy dimension, when studying the government policies toward technology, there are implications for individuals and communities, intellectual property, rights and obligations of the engineer.

Just to take an example, we could mention an increasing interest in the researches of the global catastrophic risks etc (2008).

Philosophy of engineering has a logic (methodological) dimension, too, because engineering is deeply related to methodology and requires discussions on various methodological problems, because each type of engineering has its logical particularities.

Philosophy of engineering has even a hermeneutic dimension if accepting that engineers are not as "linguistically naïve" as they are considered sometimes.

6 Horizons in the Philosophy of Engineering

A philosophy of engineering could be deployed starting from the main currents in contemporary philosophy, as observed by Carl Mitcham and Robert Mackay (2007):

> in beginning to think about possibilities for the philosophy of engineering it is useful to consider how this new regionalization of philosophy might take shape within different approaches to philosophy.

Based on the identification of six main currents in contemporary philosophy (phenomenological philosophy, postmodernist philosophy, analytic philosophy, linguistic philosophy, pragmatist philosophy and thomist philosophy), they are exploring the corresponding number of different approaches in the philosophy of engineering having each its own bibliographical references. (Mitcham & Mackay, 2007, 29).

As Mitcham and Mackay consider, those authors are: for the *phenomenological philosophy,* Don Ihde; for the *postmodernist philosophy,* Michel Foucault, Jean-François Lyotard, Jacques Derrida and Billy Vaugh Koen; for the *analytic philosophy,* Mario Bunge; for the *linguistic philosophy,* Ludwig Wittgenstein; for the *pragmatist philosophy,* John Dewey, Paul Durbin and Larry Hickman. For the *thomist philosophy* we could name Jason T. Eberl. (Mitcham & Mackay, 2007, 30)

Those currents could be considered as horizons in the philosophy of engineering, because they are not simple limits, but *limits that cannot be passed over* (the horizons exploration is not very cumulative).

From this perspective and resuming the ideas of Mitcham and Mackay, one could use topics such as:

1. "engineering and/or some of its diverse manifestations as phenomena call for more careful description and attention" in the phenomenological approach to the philosophy of engineering.
2. "question whether there even really is such a thing as engineering" in the postmodernist approach to the philosophy of engineering.
3. "conceptual distinctions between science and engineering, for the recognition of special forms of logic and knowledge in engineering, and for the engineering of ethics", "rethink[ing] engineering in terms broader than those of engineering itself while applying engineering methods to the rethinking of many other aspects of human experience" in the analytic philosophy of engineering.
4. describing "engineering as a particular language game" in the linguistic philosophy.
5. an epistemological and a social philosophical approach in the pragmatist philosophy of engineering.
6. an "attempt to understand engineering in relation to the standard branches of philosophy: metaphysics, epistemology, and ethics" in the thomist philosophy of engineering.

7 The Non-Generic Universality of Philosophy, Engineering and Being

For generic universality, non-generic universality is similar to what negentropy is for entropy.

Both the philosophy and the engineering are highly negentropic human endeavours, as they are interested in understanding, preserving and enriching the irreducible complexity of being.

More than twenty five years ago, starting from and continuing some philosophical developments of the Romanian philosophy, I have realized that, in order to understand *what-it-is/being* and to secure our knowledge, we are assuming a type of universality dependent on the quantitative identity (in which all x could, so have to be reduced to y).

By this we are reducing the richness of being, because we are *presupposing* that this is the only possible type of universality and that being itself exists universal only in this quantitative way.

Various philosophical traditions have understood the naïvety and the dangerous implications of such a static, frozen, inanimate perspective on what-it-is/being and reacted against it.

Non-generic universality is not the absence of generality, but the presence of a partial, insular, localized generality. It is the universality suggested by Ludwig Wittgenstein by the metaphors of rope and of the plan of a city, that one searched by Mircea Eliade beyond the *coincidentia oppositorum*, one in which beings, whatever they are or could be, are "islands in the stream" etc. (Guliciuc, 2009)

After the crashes of the belief in *Weltanschauung* and then in the *metanarratives*, philosophers have difficulties in their positioning concerning the reports with regard to universality. It is highly significant that, even though we are continuing, inertial and in a disguised way, to search for the generic universality everywhere, we understand more and more clearly that we do not have any idea about how the being really is, without the possibility to build a theory of the extended relativity of our knowledge.

Because it is focused on the study of being qua being and because the human being is a vectorial being (has a forward inside), philosophy is extracting order to build islands of sense.

Engineering has not and seems not to have these kinds of difficulties in its relationship with universality. Whether a philosopher or a scientist, an engineer is not searching for the generic universality of his (engineering) knowledge, but for the well made artefact – as a design, as a technological chain, as a product etc. - the type of universality he is working with is a non-generic one.

Because engineers do not dream to build the Universal Machine, but the best machine from their time, engineering is extracting order to build islands of order; of secure knowledge.

Philosophy needs engineering in order to secure its senses.

Engineering needs philosophy to find the sense of its secured knowledge.

8. Conclusions

The philosophy of engineering could have unexpected continuation from the pedagogy of the digital natives / digital immigrants of Marc Presnky (2009), when discussing wisdom in an engineering [technologized] world (Guliciuc, 2009b), when debating on concepts like "the cultured engineer", "the New renaissance Engineer" of the 21st century and his new skills.

Some philosophers are not very happy *nowadays, when engineers want to use philosophy as a tool* (Goldberg, 2007). *From the perspective here assumed – in which both philosophy and engineering are human activities with common roots in the practice, focused on the extraction and building of islands of order (complexity) – there is no problem here, because philosophy, so focused on those artefacts named* words *is a tool* (how unexpected would that be).

Finally, philosophy could be considered as the engineering of concepts, meanwhile *engineering could be considered as the search for wisdom in the making.*

THEORIES AND SPECIFICATIONS

RAYMOND TURNER

Introduction

The discovery of nature's *theories* and *laws* is taken to be one of the central goals of science. Newton's laws of motion, the laws of quantum mechanics, Einstein's theories of relativity, Maxwell's laws of electromagnetism and Darwin's theory of evolution are some of the most famous. These are big examples that are often associated with large paradigm shifts. But there are more local ones such as Hooke's law and The Huygens–Fresnel principle. However, whatever their fame and generality, all such theories form the backbone of contemporary science; their formulation, exploration and evaluation furnishes the content of much of the activity of their respective scientific disciplines. Most significantly, scientific theories are meant to be true; they are meant to express true propositional knowledge.

In contrast, engineers deal with artefacts that are *specified*, *designed, constructed* and *tested*. Examples include the Severn Bridge, the Burj Dubai tower in the Gulf emirate of Dubai and the supercomputer Blue Gene. Again these are major examples, but there are more mundane ones that include can openers and off-the-road bicycles. Even these involve engineering design activity: requirements have to be elicited, specifications presented and approved, designs created, implemented and tested. So in contrast to science, where *theories, laws and experiments* form the heart of matters, here it is *specifications, designs and artefacts.* Moreover, in contrast to theories, the latter are not things that are meant to be true; they are required to do what they have been designed for. These differing aims of science and engineering are summarised by McCarthy (2006).

There are numerous views about what (if anything) distinguishes science and engineering. One distinction is that science aims to build theories that are true, while engineering aims to make things that work.

There seems little doubt that scientists aim to construct true theories and that engineers aim to build things that work. But while the nature of truth for scientific theories forms one of the central questions in the philosophy of science, the nature of artefacts, and what it is for them to work, has been the subject of much less philosophical analysis (Fransen, 2009). The aim of this paper is to take a few tentative steps in this direction. On the face of it, theories say something about the world and specifications dictate the form of constructed artefacts. To unpack this observation we shall point out some parallels and differences between the *theory-observation* pairing and the *specification-artefact* one. In doing so, we aim to highlight some of the philosophically significant differences between the two disciplines.

Theories

Darwin's Theory of Evolution holds that all creatures naturally evolve from less complex ancestors. Random genetic mutations occur within an organism's genetic code. Some mutations, the ones that aid survival, are passed on to the next generation. Over time these accumulate and eventually evolve into different organisms. This is the process of *Natural selection*; the core of the theory of evolution (Gould, 2002; Smith, 1975).

The Special Theory of Relativity (Einstein, 1905; Das, 1993) assumes that the laws of physics are the same for all observers in uniform motion relative to one another. It further assumes that the speed of light is the same for all observers. The General Theory subjects the Newtonian Euclidian geometry of space to the maxims of special relativity. Pictorially, the dimensions of space and time are depicted as a two-dimensional surface in which massive objects create valleys and dips.

These theories are meant to be *true*. They are meant to be descriptive of nature and the way it functions. They are meant to be explanatory (Achinstein, 1983) and tell us why things are the way they are. They are intended to convey propositional knowledge about the world. Moreover, their predictions may be empirically verified. For example, general relativity predicts that light rays will bend when passing near to the sun, while evolution informs us that life is continually evolving from simple to complex organisms.

As a consequence of their descriptive and predicative powers, theories can be evaluated for their truth. While such evaluation may be a very complex process, with both the theory under evaluation and the encompassing intellectual paradigm influencing the design and choice of experiments and observations, at some level, scientific theories make predictions that can be empirically assessed, and assessment may result in the theory being accepted or rejected.

Specifications

In contrast, engineers deal with constructed artefacts that are designed and built to meet some specification. Generally, there are several stages to the design process. While different authors carve up matters differently (Hyman, 1998; Dym, 2009; Ertas, 1998; Pfleeger, 2009), there are always aspects of the following stages.

1. Problem stated
2. Designs presented
3. Artefact constructed
4. Artefact tested

To gain a little more insight into the *specification-artefact* relationship we shall consider the main conceptual content of these stages.

Two possible engineering problems are given below.

1. An incubator to be used in under-equipped medical clinics in developing countries. The incubator will maintain samples at temperatures (35° - 37° C). It will have a capacity of at least twenty 10-ml test-tubes or six 100-mm Petri dishes. The cost will be less than £100.

2. A computer store is to have named locations that hold numerical values. There has to be some means of obtaining the content of any given named location and a means of changing its contents.

These provide *specifications* of two desired artefacts. They list some of the attributes that are demanded of those artefacts; they constrain the desired artefact: an engineer, who delivered an artefact that satisfied the requirements of *1*, when asked for those of *2*, would be fired. So unlike scientific theories that have a descriptive function, specifications tell us what needs to be built. Of course, *1*, and *2* leave much unsaid. For example, for how long is the incubator to maintain the required temperatures and how are the solutions to be influenced by cost? Moreover, the second is a very different form of specification with a rather different kind of imprecision. When the location's contents are changed, and the contents of another location are abstracted, presumably its contents should not be changed. This is not explicitly said. But it can be. In particular, we might provide a mathematical description of the store and its associated operations (call them ***Update*** and ***Lookup***). According to our informal remarks these should be governed by the following conditions:

I. $Lookup(Update(s, x, v), x) = v$

II. $Lookup(Update(s, x, v), y) = Lookup(s, y) \quad where\ x \neq y$

This is an abstract mathematical description of the properties or requirements of the proposed physical device. Clearly not all specifications are amenable to such exact requirements.

Once the design problem has been specified, the engineer will turn to creating *design solutions* i.e. *designs* for the devices that meet the agreed requirements. These provide detailed plans for the construction of the artefact and, in addition, often provide a more demanding specification. Moreover, at some stage a single design solution needs to be selected and fixed[1]. This is then what forms the basis (of a more detailed specification) for the construction of the artefact.

While specifications may well have a similar logical form to theories, they have an entirely different function: they are not intended to be descriptive or explanatory. It should now be clear that their central function is *normative*. It is specifications that tell us what it is for an artefact to work. They tell us what to build and whether we have done it correctly. They influence the choice of tests that need to be done to determine whether the artefact satisfies its specification and design. For example, a bridge designed to support a given load must support that load, and experiments and tests must be designed to judge matters. If it fails the tests, the bridge is not fit for purpose. In other words, specifications furnish us with a notion of *correctness*. If we build a computer store instead of an incubator when asked for the latter, then we will not have satisfied the specification. Our artefact does not provide a *correct* implementation of the specification.

In that it is presented as a linear progression from specification to artefact, this is a rather simplified picture of engineering practice. In

---

[1] At some point in the project, freeze the design. This rule of thumb recognizes that a point is often reached in design where the character of a project, and hence the appropriate allocation of resources, changes from seeking alternative solutions to perfecting a chosen solution. As might be expected, this point is located heuristically by a trade-off between the relative risk and benefit of seeking yet another alternative. After this point is reached, a major design change runs an unacceptable risk of introducing a fatal flaw because insufficient resources remain to evaluate all of its ramifications. Once a design has been frozen (as a good rule of thumb, about 75 percent of the way into the project), the members of the design team take the general attitude, let's go with it (Kohn, 2003).

reality all kinds of reviews and backtracks will occur. Prototypes will be built and designs altered and doctored. Even the original specification may be modified. Of course, there are limits to such modification; whatever changes are made to the specification it will never be the case that a specification of a bridge will get massaged into a specification of a can opener. If the functional requirements change too much, the identity of the whole project is compromised. Indeed, these complexities and subtleties of the design process do not negate the normative function of specification. The fundamental philosophical role of specification is not tied to the actual process of construction: while the practical role of specification may be to guide the construction, its philosophical one is to provide a criterion of correctness. If it says *beans* on the can, I expect it to contain beans not bananas. How they got there may explain how the mistake was made, but not why it is a mistake.

The Scientific Stance

By and large, engineering deals with constructed artefacts and science with naturally occurring phenomena[2]. But this difference does not correspond to the differing methodologies of the two activities.

In particular, scientific theorising need not be about natural phenomena. One may take what might be broadly described as a scientific approach or stance towards engineered artefacts. More explicitly, such artefacts may be explored scientifically. Here the physical device is the thing that is given to us. It might be known to be a constructed object, but we may not understand how it works. Consequently, we might well take a naturalistic stance towards it and treat it as a natural artefact. The scientific task is to figure out what it is and what it does.

For example, suppose that we have encountered a physical device that is a large white box with small individual black windows. The windows are named with letters of the alphabet. There is small attached keyboard with letters on the upper case keys and numbers on the lower case ones. We press the keys and carry out some simple experiments. We notice that pressing the upper case letter above a window displays a number behind the window. Further exploration reveals that pressing an upper case key immediately followed by a number on the lower case keys, changes the content of the window named by the letter. Consequently, we postulate the idea that the whole device is a simple computer store. We might even formulate a theory of the device which might take the form of a theoretical

[2] Even this distinction is in need of some further analysis.

version of our specification i.e. we postulate the idea that the whole device it is a simple store with *Update* and *Lookup* operations. In other words, on finding this device, we attempt to construct some kind of *theory* about the machine i.e., we postulate an abstract machine as a theory of the physical artefact. We might even formulate axioms *I* and *II* as the theory of the latter. On this perspective, these operations are determined by what the physical machine actually does. For instance, if when given *y* and *6*, the physical machine generates the following state.

x	*y*	*z*	*w*....
5	7	9	7......

We might then postulate not *I* and *II* as its theory but the following equations:

III. $Lookup(Update(s,x,v),x) = v + 1$

IV. $Lookup(Update(s,x,v),y) = Lookup(s,y) \quad where\ x \neq y$

On this perspective, matters are determined by what the physical machine actually does (Kripke, 1982). This is the scientific approach to such artefacts i.e. methodologically, this is science, not engineering. The simple conditions *III* and *IV* do not constitute a specification of a device, but act as the basis of a scientific theory about one. They are treated as something to be judged by observations carried out on the physical artefact. They do not have normative force but descriptive content.

The Engineering Stance

On the other hand, naturally occurring artefacts such as noses, ears and lungs, can be engineered. In which case, we will use the relevant scientific knowledge to specify and design the artificial one. While this is so for most design projects, here we are specifically trying to engineer a naturalistic artefact. We may, for example, wish to reproduce a dog's nose. We shall use any scientific theories about noses and sniffing in order to inform our specification and the design of the artificial nose. The specification will have to abstract away from the physical nose, and maybe compromise on its sniffing power. A dog's nose has something like twenty times as many primary receptor cells as the human nose, and different regions of the mucous lining within the nose have different chemical properties. Unfortunately, for our engineering task, canine noses are still a bit of a mystery. And so we may have to settle for a child's nose: its power may be restricted to its ability to sense breast milk.

Indeed, such engineering may even involve an initial scientific phase where, in order to discover how it works, the natural version of the desired artefact is treated scientifically. For example, we might build scientific theories of dog's noses in order to eventually engineer an artificial one. Using the theory we somehow write a specification of the required artificial nose. Once in place, the artificial nose is the intended artefact and its success is measured by how well it meets this specification.

Of course, we might be building the artificial nose for purely scientific purposes, i.e. to explore how noses work. Once built it may function as a physical model of the dog's nose. It now takes on a different persona as a scientific model of the natural artefact which is the dog's nose. It is now evaluated against the dog's nose, not against the specification.

Theories and Specifications

These simple observations tease out some of the differences between what we might call the scientific versus the engineering perspectives. For one thing, the major differences are not located in the artefacts or phenomena. Instead, our intentional stance determines how we treat them. Whatever they are made of, whether they are abstract or concrete, engineers and scientists treat them in different ways.

The scientist will approach any object with the intention of understanding it via the construction of a theory about it. Theories are postulated and tested and amended until some kind of stability occurs. Such theories, and any surrounding assumptions and theories, are always up for revision.

In contrast, engineers design artefacts where specifications act as normative guides to their construction. The eventual specifications are not intended to constitute a theory of any existing object, but a design for new one. There is an asymmetry here. In the *theory-observation* relationship the observations are used to evaluate the theory, whereas in the *specification-artefact* one, the specification is the measure of the artefact. Testing is employed to evaluate how well the artefact measures up to its specification. The specification may well dictate the form of the tests, but it is the artefact that takes the blame when there is disagreement. In contrast, in the *theory-observation* pairing, while the theory still dictates the form of experimentation, it also takes the blame when matters go awry.

These clean characterisations offer us a bold account of the differences between the two enterprises. But it must be said that they obscure some rather difficult philosophical concerns. In particular, we have not said much about how theories are evaluated, and whether these considerations have parallels in engineering design.

Theory Correctness

The epistemological status of scientific theories is one of the central problems in the philosophy of science (Chalmers, 1999; Rosenberg, 2000). It is not our intention to get involved in the debate over the nature of scientific knowledge. We shall only sketch some of the moves in the overall drama in order to see how parallel ones play out in the engineering enterprise.

The problem of induction is one of the core themes in the philosophy of science. It throws doubt on the claim that a theory can be logically confirmed. Scientific laws may involve quantification over potentially infinite domains. For example, the equations below quantify over real numbers:

$$\mathbf{F=Ma}$$
$$\mathbf{E=Mc^2}$$

And so they cannot be logically deduced from a finite number of observation statements.

Popper (1972) reacted to this problem by denying that the aim of science was theory confirmation. Instead its aim is linked to theory falsification; and it is the ability to be falsified by experiment that distinguishes science from non science. While general laws cannot be deduced from a finite number of observation statements, they can be falsified by one negative one.

Kuhn (1970) attacked this conception of scientific progress. Science does not proceed in this clean-cut way with theories being postulated and falsified. Moreover, a clash between a theory *N* and observation statement *O* may not necessarily mean that the theory *N* is false. While one black swan may falsify the general law that all swans are white, actual scientific instances are generally quite complex. Observations are not made in a theory independent way, and theories come embedded in layers of assumptions, including higher level theories and paradigms. These background packages make it difficult to know what to blame. The interpretation of the observation statement may be filtered by the whole background package. So deciding which aspects of a theory and its theoretical surroundings need to be revised, including the force to be attached to the observation statement itself, is often a delicate matter.

Lakatos (1980) attempted to restore sanity by dividing a theory/paradigm into a core theory that in the normal practice of science is not questioned, and a further class of assumptions that supplies a surrounding umbrella of protection. The latter might be tampered with as long as doing so results in novel and testable predictions.

This is a very crude and informal account of how the philosophical literature portrays the nature of scientific knowledge. We mention these matters only to illustrate that the *theory-observation* relationship is far from simple. Finding an observation that conflicts with a theory does not immediately mean that it is the theory that must be given up. Nevertheless, whatever the difficulties associated with such evaluation, it is the theory that is under scrutiny.

The Correctness of Artefacts

In contrast, in engineering it is the artefact: it has to work. To decide matters one has to devise tests to check whether the artefact behaves correctly, and this is determined by its specification. Does it behave like a computer store: does it store values? Does it behave like an incubator: does it maintain samples at required temperatures?

This is the opposite picture to the *theory-observation* relationship. While in the latter it is the theory that needs to be measured against the world of observations, here it is the artefact that is to be measured against the criteria laid down by the specification. Presumably, the latter is an abstract notion, and the artefact may be physical or abstract. For example, a program specification may be taken to be a specification of a program in a programming language or a digitised image on a CD. Is the *specification-artefact* relationship less problematic than the *theory-observation* one?

We shall investigate matters by reference to software engineering (Pfleeger, 2009). A team of programmers have built an implementation of a given specification, and we have been assigned the task of showing that it meets its specification. We have been given the program source code and a copy of the program burnt onto a CD. There are two possible tasks here which correspond with the two possible items that may be taken to be the designed artefact: both the source code and the digital object that is burnt onto the CD may be so considered. What constitutes correctness for each of these two artefacts is different (Fetzer, 1988; Colburn, 2006). The former is subject to formal analysis where the specification and the program are treated as mathematical objects, and the correctness of the software is mathematically established. Here we are not concerned with this mathematical stage; we shall assume that this proof is in place. We shall reflect upon the correctness of the physical artefact that occupies the CD. This is a very different issue.

This is not a mathematical notion of correctness. Indeed, here we have problems that parallel the evaluation of scientific theories. Presumably, we test the digital device on some stock examples. These are usually chosen to cover some standard cases, and to provide confidence in the physical device. We may find that it satisfies the specification in all the chosen cases. What does this demonstrate? As with the case of scientific theories, it does not guarantee that the digital mechanism works for all cases: we cannot logically conclude that it will always work on the basis of a finite number of tests, no matter how carefully chosen. In general, correctness statements will involve quantification over an unbounded number of possibilities. To test every possible case would involve running the software on every possible combination of parameter values. Even if such testing was limited to actual machines, in general, it would be practically infeasible.

However, there is an important twist compared with the evaluation problem for scientific theories. In engineering design it is the artefact not the abstract specification that is being tested. Nor is it a scientific theory that is subject to induction worries, but a physical artefact that cannot be proven to meet its abstract specification.

Much the same fate applies to the falsification approach. If the physical version of the software fails one test, we cannot conclude that the physical version does not satisfy its specification. The digital version of the software is running on a very complex infrastructure that is made up of layers of digitised software and physical machines. It could be any part of this complex system that is malfunctioning. The best we can do is to accumulate inductive evidence as to the source of the failure. It cannot be pinned down to the software that is on the CD.

No doubt, in order to overcome these difficulties, one might attempt to take the analysis further and mimic some of the moves made in Lakatos (1980). However, we have done enough to indicate that there are reasons to think that the correctness of engineering artefacts raises parallel philosophical concerns to those associated with scientific theories.

Finally, observe that on this characterisation, software development is not a scientific endeavour (Colburn, 2006). Software specifications are not scientific theories, and it is artefacts not theories that are tested.

It seems that the truth of theories and the correctness of artefacts breed similar epistemological problems. Of course, in practice, judgements are made about the success or otherwise of an engineered artefact. However, epistemological questions are concerned with the clarification and the philosophical basis of such judgements. And here it seems that the *theory-observation* and the *specification-artefact* pairings face parallel epistemological worries. So when McCarthy (2006) claims that engineering knowledge is somehow more certain than scientific knowledge, it is not clear that the case has been made.

MORAL MEDIATORS IN TECHNOLOGY

LORENZO MAGNANI

1. Rational Acting in a Human Unsettled World

Morality could be defined, at the very last, as

> the effort to guide one's conduct by reason - that is, to do what there are the best reasons for doing - while giving equal weight to the interests of each individual who will be affected by one's conduct: there are not privileged people. (Rachels, 1999, p.19).

Moral reasoning could be viewed as a form of "possible worlds" anticipation, a way of getting chances to shape the human world and act in it. It could be of help to prefigure risks, possibilities, and effects of human acting, and to promote or prevent a broad variety of guidelines. Hence, we need 1) to possess good and sound principles/reasons applicable to the various problems, able to give rise to arguments that can be offered for opposite moral views, and 2) appropriate ways of reasoning which permits us to apply the available reasons in the best way. "Creating ethics" means creating the world and its directions, in front of different (real or abstract) situations and problems. This process requires the adoption of skilful and creative ideas, in order to react in response to new, previously unknown cases, or in cases of moral conflict. In this way events and situations can be reinvented either as an opportunity or as a risk for new moral directions.

2. Respecting Things as People, Respecting People as Things

It is well-known that Immanuel Kant's categorical imperative states, in *Groundwork of the Metaphysics of Morals*:

> *Act only on that maxim through which you can at the same time will that it should become a universal law.* (Kant, 1964, 88)

When dealing with "The formula of the end in itself" (*Groundwork*, 88) Kant observes that

> [...] man, and in general every rational being *exists* as an end in himself and not merely as a means for arbitrary use by this or that will: he must in all his actions, whether they are directed to himself or to other rational beings, always be viewed *at the same time as an end.*
> [...] Beings whose existence depends, not on our will, but on nature, have none the less, if they are not rational beings, only a relative value as means and are consequently called *things*. Rational beings, on the other hand, are called *persons* because their nature already marks them out as ends in themselves – that is, as something which ought not to be used merely as a means – and consequently imposes to that extent a limit on all arbitrary treatment of them (and is an object of reverence). [...] Persons, therefore, are not merely subjective ends whose existence as an object of our actions has a value *for us*; they are *objective ends*, that is, things whose existence is in itself an end, and indeed an end such that in its place we can put no other end to which they should serve *simply* as means. (*Groundwork*, 95-96)

Kant uses the word "end" in a very formal way, as synonymous with "dignity"; its teleological nature is, after all, not important. Kant is very clear on this point when he writes that

> Teleology views nature as a kingdom of ends; ethics views a possible kingdom of ends as a kingdom of nature. In the first case the kingdom of ends is a theoretical Idea used to explain what exists. In the second case it is a practical Idea used to bring into existence what does not exist but can be made actual by our conduct – and indeed to bring it into existence in conformity with this Idea. (*Groundwork*, 104)

Hence, Kant defines the "kingdom" as a "systematic union of different rational beings under common laws" (*Groundwork*, 104).

Kant's considerations lead us to the following practical imperative:

> Act in such a way that you always treat humanity, whether in your own person or in the person of any other, never simply as a means, but always at the same time as an end. (*Groundwork*, 96)

In the

> kingdom of ends everything has either a *price* or a *dignity*. If it has a price, something else can be put in its place as an *equivalent*; if it is exalted

> above all price and so admits of no equivalent, then it has a dignity. (*Groundwork*, 101)

Things that human beings need have a "market price"; moreover, items that are merely desired rather than needed have an affective "fancy price" [*Affektionspreis*]. But

> [...] that which constitutes the sole condition under which anything can be an end in itself has not merely a relative value – that is, a price – but has an intrinsic value – that is, *dignity*. (*Groundwork*, 101)

A simple example that illuminates the Kantian perspective relates to human moral behaviour and the issue of responsibility. Economists say that a decision results in a negative externality when someone other than the decision maker ends up bearing some of the decision's cost. Responsibility is externalized when people do not take responsibility for the problems they cause and delegate finding a solution to someone who had no part in creating the trouble. When those who must deal with the consequences of the decision are not aware such a task has been delegated to them, they are treated as means. Of course, on the other hand, responsibility is internalized when people accept responsibility for the outcome of their actions.

Kant's wonderful lesson can be inverted: it is possible for things to be treated or respected in ways one usually reserves for human beings. Many things, or means, previously devoid of value, or previously valuable only in terms of their market price or affective price, can also acquire a moral status or intrinsic value. Conversely, just as things can be assigned new kinds of value, so too can human beings, for there are moral positive aspects of treating people like things, as we shall see.

Anthropocentric ideas, like those that inform Kant's imperative, have made it difficult for people to acquire moral values usually associated with things and for things to attain moral worth traditionally reserved for people. We said that, in Kantian terms, people do not have to be "treated as means (and only as means)." I propose upgrading that idea with a new one – respecting people as things in a positive sense. In this scenario, people are respected as "means" in a way that creates a virtuous circle, one in which positive moral aspects enjoyed by things can be used to reshape moral endowments attributed to things.

Attributing moral worth to things can be seen as a combination of the Kantian imperative and of John Stuart Mill's idea of freedom:

> The only freedom which deserves the name is that of pursuing our own good in our own way, so long as we do not attempt to deprive others of theirs, or impede their efforts to obtain it. (Mill, 1966, 18)

If, as Mill teaches, beings (or things, we now add) have the right to something, they are entitled not only to the goal itself but also to the unobstructed pursuit of it. When things also became regarded as entities with interests and rights of their own, the philosophical conceptual space of utilitarianism (animals suffer!) and the idea of environmental ecology were constructed. How did this happen?

A special kind of "things", namely animals, has always been used to lend various aspects of their own properties and functions to human beings, for instance in biomedical research. In this field *animal models* have served to induce certain conditions in animals for deriving conclusions about some conditions in human beings. The results are achieved by exploiting analogies (for instance rats and humans are alike in various ways) rather than disanalogies. This theme is very important in philosophy of science, because modelling is widespread in scientific practice. Many epistemological problems arise, like the quest of the qualities that make an animal model valid and appropriate (Shelley 2004). I contend that also in the area of ethics we have to look at the moral "models" that come from things, like animals and objects: as I already said, people can be respected as "means" in a way that creates a virtuous circle, one in which positive moral aspects enjoyed by things can be used to reshape moral endowments attributed to humans.

Perhaps the first "things" to gain new moral rights in western culture were women, a change that was not universally welcomed. Indeed, the ideas propagated in this direction by Mary Wollstonecraft in her 1792 treatise *A Vindication of the Rights of Women* were initially considered absurd (Singer, 1974). This sort of ideological conflict has been played out again in the last few decades as animal rights advocates and environmental ethicists have waged a struggle similar to the one women faced in the eighteenth century – that of redefining a means as an end. To achieve that goal, some intellectuals and activists have sought to reframe how various plants, animals, ecosystems – even the land itself – are valued so that they are regarded as "ends" and accorded the rights and protection that status entails.

3. Human and Non-Human Collectives

In *Pandora's Hope*, Bruno Latour observes that humans and non-humans are inextricably intertwined:

> You are different with a gun in your hand; the gun is different with you holding it. (Latour, 1999, 179)

We are in some sense "folded" into non-humans, so that we delegate action to external things (objects, tools, artefacts) that in turn share our human existence with us. The idea of the "collective" expresses an exchange of human and non-human properties akin to what I have just described in the case of things in search of intrinsic values:

> What the modernist science warriors see as a horror to be avoided at all costs – the mixing up of objectivity and subjectivity – is for us, on the contrary, the hall mark of a civilized life. (*Pandora's Hope*, 200)

Many such examples are mentioned by Bruno Latour: using knowledge about non-humans to reconfigure people and, conversely, projecting on non-humans the properties and functions of humankind. When considered from the ethical perspective, the first case depicts our problem of respecting people as things, while the second depicts ideas illustrated earlier about the moral representations of non-humans:

> The new hybrid remains a non-human, but not only has it lost its material and objective character, it has acquired properties of citizenship. (*Pandora's Hope*, 202)

Of course, the non-moral case of endowing non-humans entities with speech, intelligence, and other human properties – things from classical media to computational tools, from paintings to artificial intelligence, from simple tools like a hammer to sophisticated machines – is related to this movement. So, too, are agriculture and the domestication of non-human animals, which involves their socialization and re-education.

In turn, external things (electrical, transportation, and telecommunications industries, for example) have constructed new social frameworks for people, and so in the case of the many new roles delineated by factories, machines, and institutions in establishing new constraints in managing humans and in stabilizing new types of human negotiations:

> It was from techniques, that is, the ability to nest several subprograms, that we learned what it means to subsist and expand, to accept a role and discharge a function. (*Pandora's Hope*, 209)

Tools, that have always played the role of human prostheses, become integrated into our bodies as we use them in a kind of anthropological transformation of both the individual and the collectives. This mixture between human and non-human is also expressed in human bodies that are increasingly shaped and integrated by "sociotechnical negotiations and artefacts."

The cyclical process of transferring qualities between humans and non-humans is, of course, an inextricable part of our evolution, and consequently it requires ongoing negotiations and a continual redrawing of the lines between the two kinds of entities.

4. Hybrid People

Following Andy Clark's conclusions on the relationships between humans and technology, we all are "constitutively" natural-born cyborgs, that is, biotechnologically hybrid minds (Clark, 2003). Less and less are our minds considered to be in our heads: human beings have solved their problems of survival and reproduction by "distributing" cognitive functions to external non-biological sources, props, and aids. Our biological brains have delegated to external tools many activities that involve complex planning and elaborate assessments of consequences (*Natural-Born Cyborgs*). A simple example might be how the brain, when faced with multiplying large numbers, learns to act in concert with pen and paper, storing part of the process and the results outside itself. The same occurred when Greek geometers discovered new properties and theorems of geometry: they manipulated external diagrams to establish a kind of continuous cognitive negotiation with a suitable external support (like sand or a blackboard), to gain new important information and heuristic suggestions. The use of external tools and artefacts is very common: cognitive skills and performances are so widespread that they become invisible, thus giving birth to something I have called "tacit templates" of behaviour that blend "internal" and " external" cognitive aspects.[1]

New technologies will facilitate this process in a new way: on a daily basis, people are linked to non-biological, more-or-less intelligent machines and tools like cell phones, laptops, and medical prosthetics. Consequently, it becomes harder and harder to say where the world stops and the person begins. Clark contends that this line between biological self and technological world has always been flexible and that this fact has to be acknowledged both from the epistemological and the ontological points of view. Thus the study of the new anthropology of hybrid people

[1] Tacit templates of moral behavior in relation to moral mediators are treated in Magnani 2007, chapter six. Their epistemological counterpart, which has to do with manipulative abduction, is illustrated in the same book, chapter seven.

becomes important, and I would add that it is also critical for us to delineate and articulate the related ethical issues. Some moral considerations are mentioned in the last chapter of Clark's book, in which he addresses important issues such as inequality, intrusion, uncontrollability, overload, alienation, narrowing, deceit, degradation, and disembodiment – topics that are especially compelling given recent electronic and biotechnological transformations. Nevertheless Clark's approach does not shed sufficient light on basic ethical problems related to identity, responsibility, freedom, and control of one's destiny; problems that accompany technological transformations. He clearly acknowledges such issues, but only in a minimal and general way:

> Our redesigned minds will be distinguished by a better and more sensitive understanding of the self, of control, of the importance of the body, and of the systemic tentacles that bind brain, body, and technology into a single adaptive unit. This potential, I believe, far, far overweighs the attendant threats of desensitization, overload, and confusion. (*Natural-Born Cyborgs*, 179) […]. Deceit, misinformation, truth, exploration, and personal reinvention: the Internet provides for them all. As always, it is up to us, as scientists and as citizens, to guard against the worst and to create the culture and conditions to favor the best. (*Natural-Born Cyborgs*, 187)

As I contend in my book, I think these problems are more complicated, and teasing out their philosophical features will require deeper analyses. What new knowledge must we build to meet the challenges of living as hybrid people? I certainly share Clark's enthusiasm in philosophically acknowledging our status as "cyborgs," but I would like to go further, to do more than just peer through the window of his book at the many cyberartefacts that render human creatures the consumers-cyborgs we are.

Our bodies and our "selves" are materially and cognitively "extended," meshed, that is, with external artefacts and objects, and this fact sets the stage for a variety of new moral questions. For example, because so many aspects of human beings are now simulated in or replaced by things in an external environment, new ontologies can be constituted – and Clark would agree with me. Pieces of information that can be carried in any physical medium are called "memes" by Richard Dawkins (1989). They can "stay" in human brains or jump from brain to brain to objects, becoming configurations of artificial things that express meaning, like words written on a blackboard or data stored on a CD, icons and diagrams on a newspaper, configurations of external things that express meaning like an obligatory route. They can also exist in natural objects endowed with informative significance – stars, for example, which offer navigational guidance. In my perspective the externalization of these chunks of information is described in the light of the cognitive delegation human beings concentrate in material objects and structures.[2]

Beyond the supports of paper, telephone, and media, many human interactions are strongly mediated (and potentially recorded) through the Internet. What about the concept of identity, so connected to the concept of freedom? At present identity has to be considered in a broad sense: the amount of externally stored data, information, images, and texts that concern us as individuals is enormous. This storage of information creates for each person a kind of external "data shadow" that, together with the biological body, forms a "cyborg" of both flesh and electronic data that identifies us or potentially identifies us. I contend that this complex new "information being" depicts new ontologies that in turn involve new moral problems. We can no longer apply old moral rules and old-fashioned arguments to beings that are at the same time biological (concrete) and virtual, situated in a three-dimensional local space but potentially "globally omnipresent" as information-packets. For instance, where we are located cybernetically is no longer simple to define, and the increase in telepresence technologies will further affect this point. It becomes clear that external, non biological resources contribute to our variable sense of who and what we are and what we can do.

For now, however, let us reconsider the consequences of my motto "respecting people as things" when applied to the idea of hybrid people.

5. Distributed Morality and Technology

I call the external objects and structures in science, to which cognitive aspects and roles are delegated *epistemic mediators*, a blackboard with a diagram, for example (Magnani, 2007, chapter seven). In my book on creative reasoning, I have described epistemic mediators not only as external objects and structures but also as human organizations – in this case, viewed as distributors of externalized cognitive potentialities (Magnani, 2001). Cognitive mediators function as enormous new external sources of information and knowledge, and, therefore, they offer ways of managing objects and information that cannot be immediately represented or found internally using only "mental" resources. Analyzing these external

---

[2] I addressed the role of this kind of cognitive delegation from an ethical perspective in the sections "Being Moral through Doing: Taking Care" (chapter six) and "The Logical Structure of Reasons" (chapter seven) and from an epistemological perspective in "Cognitive and Epistemic Mediators" (chapter seven) of my book (Magnani, 2007).

structures is especially important in clarifying the role of media and of computational and information techniques. Epistemic mediators also help to organize social and cognitive decisions made in academic settings: examples of epistemic mediators are for instance artefacts in a scientific laboratory (a telescope or a magnetic resonance imaging machine) but also the organized collective of scientists itself, that is characterized by a specific distribution of cognitive roles, skills, and duties.

I think the best approach to studying these problems is to use so-called *computational philosophy*. The advent of certain machines and various rational methods and models brought about a computational turn in the last century, and this shift has revealed new ways to increase knowledge by embedding it in scientific and technological environments and by reshaping its major traditional topics (Magnani, 1997). Just to make an example, the role of PCs and the Internet in improving scientific research is very clear. In the new century, computational philosophy will allow an analysis of problems in recent logical, epistemological, and cognitive aspects of modelling activities employed in scientific and technological discovery. Computational philosophy supplies modern tools (new concepts, methods, computational programs and devices, logical models, etc.) to reframe many kinds of cultural (philosophical, ethical, artistic, etc.) knowledge that would remain inaccessible using old approaches, just mainly centred on the exploitation of mere "narratives".

It is in this intellectual light that I introduce the concept of the *moral mediator*. Moral mediators play an important role in reshaping the ethical worth of human beings and collectives and, at the same time, facilitate a continuous reconfiguration of social orders geared toward rebuilding new moral perspectives. Finally, thinking in terms of cognitive capacities, a human being can be considered a kind of "thing" that can incorporate information, knowledge, know-how, cultural tradition, etc., just as cognitive objects like a book, a PC, or a work of art do. Unfortunately, human beings are sometimes assigned less value than things.

6. Delegating Ethics and the Role of Moral Mediators

6.1. Templates of Moral Doing

It is difficult to establish an exhaustive list of invariant behaviours that can be considered ethical manipulative reasoning. As illustrated above, expertly manipulating non-human objects in real or artificial environments requires old and new *templates* of behaviour that are repeated at least somewhat regularly. Only exceptionally we are referring here to action that simply follows articulated, previously established plans; at issue are embodied, implicit patterns of behaviour that I call tacit templates. This variety of "hidden" moral activity is still conjectural: these templates are embedded moral hypotheses that inform both new and routine behaviours, and, as such, enable a kind of moral "doing." In some situations, templates of action can be *selected* from those already stored in the mind-body system, as when a young boy notices his baby sister crying and, without thinking, automatically tries to comfort the infant by stroking her head or singing a lullaby as he has seen his parents do many times. In other instances, new templates must be *created* in order to achieve certain moral outcomes. Such newly forged behaviour patterns are, as we will see, important components of the concept of knowledge as a duty. New challenges require new templates and in my book (Magnani, 2007) I have illustrated many new challenges generated for example by technological products.

The following tacit templates of moral behaviour present interesting features:

1. sensitivity to *curious or anomalous aspects* of the moral situation. In this case manipulations are performed to reveal potential inconsistencies in received knowledge, as when we suddenly adopt a different embodied attitude toward our spouses to elicit a reaction that confirms or discounts hypotheses about their feelings or to develop new hypotheses about the relationships This might be the case when a man becomes more aggressive to check his wife's tolerance and caring for him. Or when investigating a crime, detectives spontaneously further investigate the evidence to get more interesting data to build a moral data shape of the suspect;

2. preliminary sensitivity to *dynamic character* of the moral situation, and not only to entities and their properties. A common aim of manipulations is to practically reorder the dynamic sequence of the events and of the human relationships associated with the moral problem in order to find new options for action. An example might be a woman who, having decided to have an abortion, then spontaneously tries to modify the dynamical aspects of both her behaviour and her relationships in hopes of establishing new perspectives helping her to envisage a possible decision different from the first one first envisaged. She is unconsciously changing her behaviour in hopes of making herself decide against the abortion;

3. referral to manipulations that exploit *artificial created environments and externally induced feelings* to free new possibly stable and repeatable sources of information about hidden moral knowledge and constraints. This template feature is apparent, say, in a discussion of the moral problem

of capital punishment when we exploit resources like statistics, scientific research, or information from interviews to gather real rather than faulty information, like the one about the genuine relief the murder victim's relatives feel when the criminal is killed. In this way a new configuration of the social orders of the affected groups of people is achieved;[3]

4. various contingent ways of spontaneous moral acting. This case contemplates a cluster of very common moral templates. A person will automatically look at issues from different perspectives; *assess* available information; *compare* events; *test*, *choose*, *discard* and image additional manipulations; and implicitly *evaluate* possible *new orders* and *relationships* (for instance simpler orders, to facilitate analogies or comparisons).These strategies are all useful ways to get suitable evidence to test previously established moral judgments also through stimulating the derivation of significant consequences of those judgments;[4]

More features of our tacit templates are related to the following additional issues:

5. spontaneous moral action that can be useful in presence of *incomplete or inconsistent information* or a *diminished capacity to act morally* upon the world. Such action works on more than just a "perceptual" level – it is also used to get additional data that restores coherence and/or improves deficient knowledge;

6. *action as a control of sense data* illustrates how we can change the position of our bodies (and/or of the external objects) to reconfigure social orders and collective relationships; it also shows how to exploit artificially created events to get various new kinds of stimulation. Action of this kind provides otherwise unavailable tactile, visual, kinaesthetic, sentimental, emotional, and bodily information that, for example, helps us take care of other people (cf. below in the following subsection);

7. action enables us to build new *external artefactual models* of ethical mechanisms and structures (through "institutions," for example) to substitute for the corresponding "real" and "natural" ones. (Keep in mind, of course, that these "real" and "natural" structures are also artificial – our cultural concept of "family" is not a natural institution.) For instance, we can replace the "natural" structure "family" with an environment better suited for an agent's moral needs, which occurs when, say, we remove a child from the care of abusive family members. In such a case we are exploiting the power of an *artificial* "house" to reconfigure relationships.

---

[3] On the reconfiguration of social orders realized in science (laboratories), cf. (Knorr Cetina 1999).

[4] Analogues of all these manipulative templates are active in epistemic settings: cf. (Magnani 2001).

A different setting – a new but still artificial framework – facilitates the child's recovery and allows him or her to rebuild moral perceptions damaged by the abuse. A similar effect occurs when people with addiction problems move into group homes where they receive treatment and support. An even simpler example might be the external structures we commonly use to facilitate good manners and behaviour: fences, the numbers we take while waiting at a bakery, rope-and-stanchion barriers that keep lines of people in order, etc.

Of course many of the actions that are entertained to build the artefactual models above are not tacit, but explicitly projected and planned. However, imagine the people that first created these artefacts (for instance the founders of the group houses for addicted people), it is not unlikely that they created them simply and mainly "through doing" (creation of new tacit templates of moral actions) and not by following already well-established projects. Many of the actions which are performed to build technological artefacts and machine endowed with moral delegations (moral mediators) are of this type.

6.2. Moral Agents and Moral Patients

Technological artefacts and machines are designed, produced, distributed, and understood in the human world; they are strictly intertwined with the social interactions of humans: technology affects what people do and how they do it. For example computers possess moral agency because they 1. have a kind of intentionality and 2. can have effects on the so-called "moral patients" (see below), that is they can harm or improve the interests of beings capable of having their interests impeded or furthered:

> Artefacts are intentional insofar as they are poised to behave in a certain way when given input of a particular kind. The artefact designer has a complex role here for while the designer's intentions are in the artefacts, the functionality of the artefact often goes well beyond what the designer anticipated or envisaged. Both inputs from users and outputs of the artefacts can be unanticipated, unforeseen, and harmful. (Johnson 2004)

Some ethicists maintain that entities can be framed as moral *patients* and as moral *agents*. Not only human beings but also things can be conceived of as moral patients (as entities that can be acted upon for good and evil) and also as moral agents (as entities that can perform actions and are sources of moral action, again for good or evil).

There are many cases:

1. the two classes are disjoint (no entity qualifies as both an agent and a patient, this is clearly unrealistic);
2. the first class can be a proper subset of the second;
3. and the two classes intersect each other (both cases 2. and 3. are not promising because they both require at least one moral agent that in principle could not qualify as a moral patient - we only have supernatural agents that can fulfil this requirement, for example a God that affects the world but is not affected by the world);
4. all entities that qualify as agents also qualify as patients and vice versa (standard position), and, finally,
5. all entities that qualify as patients also qualify as agents.[5]

The fact that animals seem to qualify as moral patients that are excluded from playing the role of moral agents requires a change in the perspective 5. In short, certainly "things" (and so artificial entities)[6] extend the class of entities that can be involved in a moral situation, both as moral agents (for instance Internet) and as moral patients that enjoy intrinsic values (for instance a work of art). Of course the properties enjoyed by "things" of being a moral agent or patient are not the same as that of human beings. To make an example, artefacts can be agents of moral actions, but they are neither responsible nor exhibit free will, full intentionality, and emotions like human beings.

I think this distinction between moral patients and agents is certainly correct and useful, nevertheless obliterates the dynamic aspects instead explained following my perspective in terms of moral delegation and externalization. Indeed moral delegation to external objects and artefacts does not take place because a given thing is supposed to intrinsically possess a given set of properties appraised on their own. For example, the Gioconda has no free will, no proper intentions, and so on. However, the way it dynamically interacts with humans, and how they respond to it, is what gives value to it. In this sense, my conception differs from the one that distinguishes moral patient from moral agent.

---

[5] *Cf.* Floridi and Sanders, 2004. Carstein Stahl (2004) has recently investigated the problem concerning whether computers can be considered autonomous moral agents. Since computers cannot understand the information they store and manage, they lack the basic capacity "to reflect morality in anything". He argues on this point introducing an interesting and curious test called "the moral Turing test".

[6] On the legal extension of personhood to artificial agents (for instance shopping websites), cf. the interesting conclusions of the recent Chopra and White, 2004. Very up-to-date issues related to the contracts entered into by artificial agents and to their punishment and financial penalties are also discussed.

According to that view, the Gioconda (or an Internet selling system) would be a moral patient, because it does not possess all those features shared (or supposed to be shared) by human beings (conscious will, an actual free will, proper intentions, etc.). However, this view fails to account for the process by which we continuously delegate and give (moral) value to the things that are around us. For example, how could the patient-agent distinction account for the reason why the first present you received from your girlfriend may acquire such a great (intrinsic) value? It could be an old and haggard t-shirt, but it doesn't matter, indeed.

Moreover, there is an additional reason to prefer my conception about moral delegation described above. The idea that animals should be respected, or should have rights on their own is also based on the claim that animals suffer as well as we do. They are moral patients and as patients they have to be respected.

According to my view, this is a result of a moral mediation. As we delegate to the animals new moral worth, we use them to depict previously unseen new moral features of suffering, which for human beings acquires a new value and a new extension. Animals played the role of moral mediators because they mediated new aspects of human beings' moral lives.[7]

The patient-agent distinction specially elicits differences: it is very obvious that the moral agency of computers is not the same as that of human beings, and in this respect it is not different in kind from that of other technologies. It has been argued that computers have a kind of external intentionality (that is expressed in states outside of the body, such as speech acts, written sentences, maps, and other designed artefacts), but they cannot have internal intentionality: their agency can be compared to human "surrogate" agency, such as tax accountants or estate executors (Powers, 2004). This illustrates the kind of moral character of computer systems by showing that computer systems have a kind of intentionality and have effects on moral patients, hence they are appropriate objects of moral appraisal. In these cases we are faced with a kind of "mindless morality" (Floridi and Sanders, 2003). The problem of the moral agency of artefacts also involves the construction of the suitable policies we can (and/or have to) adopt for "punishing" – that is censoring, modifying, re-engineering, removing – them.

I think the more extended concept of "moral mediator" can better encompass and explain the issues above: the moral patients and moral agents are special cases of moral mediators.

[7] I will detail this point below in the following section "Moral Mediators".

6.3. Moral Mediators

The considerations in the previous subsection indicate the fact that a significant portion of manipulations is also devoted to building a vast new source of information and knowledge: external *moral mediators*. I have derived this expression from "epistemic mediators," a phrase I introduced in a previous book (Magnani, 2001, chapter three), which consist of external representations, objects, and artefacts that are relevant in scientific discovery and reasoning processes. As I have already illustrated moral mediators represent a kind of redistribution of the moral effort through managing objects and information in such a way that we can overcome the poverty and the unsatisfactory character of the moral options immediately represented or found internally (for example principles, prototypes, etc.). I also think that the analysis of moral mediators can help accounting for the mechanisms of the

> macroscopic and growing phenomenon of global moral actions and collective responsibilities resulting from the 'invisible hand' of systemic interactions among several agents at local level. (Floridi and Sanders 2003, p. 178).

More than just a way to move the world toward desirable goals, action also serves a moral role: we have said that when people do not have adequate information or lack the capacity to act morally upon the world, they can restructure their worlds in order to simplify and solve moral tasks. Moral mediators are also used to elicit latent constraints in the human-environment system. The links discovered grant us access to precious new ethical information. For instance, let us imagine a wife whose work requires long hours away from her husband, and her frequent absences cause conflict in their relationship. She then spontaneously begins to spend more quality time with her spouse in an attempt to save their marriage. The mediating effect of her spontaneous action can cause variables affected by "unexpected" and "positive" events in the relationship to covary with informative, sentimental, sexual, emotional, and, generally speaking, bodily variables. There was no discernible connection between these hidden and overt variables before the couple adopted a reconfigured "social" order – that is, increased time together – and uncovering such links reveals important new information, which, in our example, might be renovated and unexpected sexual pleasure, astonishing intellectual agreement, or surprising identical emotional concerns on specific matters.

Natural phenomena can also serve as external artefactual moral mediators: when in previous sections we considered the problem of "respecting people as things," we were referring to the ability of external "natural" objects to create opportunities for new ethical knowledge, as in the case of endangered species: we have learned something new by examining how people seek to redefine themselves as "endangered". Many external things that have been traditionally considered morally inert can be transformed into moral mediators. For example, we can use animals to identify previously unrecognized moral features of human beings or other living creatures, as we can do with the earth, or (non natural) cultural entities; we can also use external "tools" like writing, narrative, ritual, and various kinds of pertinent institutions to reconfigure unsatisfactory social orders. Hence, not all moral tools are inside the head; many are shared and distributed in external objects and structures that function as ethical devices.

External moral mediators function as components of a memory system that crosses the boundary between person and environment. For example, they are able to transform the tasks involved in simple manipulations that promote further moral inferences at the level of model-based abduction.[8] When an abused child is moved to a house to reconfigure her social relationships this new moral mediator can help her to experience new model-based inferences – new model-based cognitive hypotheses – (for instance new emotions concerning adults and new imageries about her past abuse).

Moreover, I can alter my bodily experience of pain through action by following the template *control of sense data*, as we previously outlined, that is through shifting – unconsciously – the position of my body and changing its relationships with other humans and non-humans experiencing distress. Mother Theresa's personal moral rich feeling and consideration of pain had been certainly shaped by her closeness to starving and miserable people and by her manipulation of their bodies. In many people, moral training is often related to these kinds of spontaneous

---

[8] I introduced the concept of model-based abduction in (Magnani, 2001). The term "model-based reasoning" is used to indicate the construction and manipulation of various kinds of representations, not mainly sentential and/or formal, but mental and/or related to external mediators. Obvious examples of model-based reasoning are constructing and manipulating visual representations, thought experiment, analogical reasoning, but also emotional feeling. Of course, abductive reasoning - which is reasoning to hypotheses - can be performed in a model-based way, internally or with the help of external mediators. In this case, I am referring to an activity of producing "moral" hypotheses in an abductive model-based way.

(and "lucky") manipulations of their own bodies and sense data so that they build morality immediately and non-reflectively "through doing."

Throughout history, women have traditionally been thought to place more value on personal relationships than men do, and they have been generally regarded as more adept in situations requiring intimacy and caring. It would seem that women's basic moral orientation emphasizes taking care of both people and external things through personal, particular acts rather than relating to others through an abstract, general concern about humanity. The ethics of care does not consider the abstract "obligation" as essential; moreover, it does not require that we impartially promote the interests of everyone alike. Rather, it focuses on small-scale relationships with people and external objects, so that, for example, it is not important to "think" of helping disadvantaged children all over the world (like men aim at doing) but to "do" so when called to do so, everywhere.[9]

My philosophical and cognitive approach to moral model-based thinking and of morality "through doing" does not mean that this so-called female attitude, being more closely related to emotion, should be considered less deontological or less rational and therefore a lower form of moral expression. I contend that many of us can become more intuitive, loving parents and, in certain situations, learn to privilege the "taking care" of our children by educating our feelings – maybe by heeding "Kantian" rules.[10] The route from reason to feeling (and, of course, from feeling to reason) is continuous in ethics. Many people are suspicious of moral emotional evaluations because emotions are vulnerable to personal and contextual attributes. Nevertheless, there are moral circumstances that require at least partially emotional evaluations, which become particularly useful when combined with intellectual (Kantian) aspects of morality.

Consequently, "taking care" is an important way to look at people and objects and, as a form of morality accomplished "through doing," achieves status as a fundamental kind of moral inference and knowledge. Respecting people as things is a natural extension of the ethics of care; a person who treats "non-human" household objects with solicitude, for example, is more likely to be seen as someone who will treat human

---

[9] Moreover, both feminist skepticism in ethics and the so-called "expressive-collaborative model" of morality look at moral life as "a continuing negotiation *among* people, a socially situated practice of *mutually* allotting, assuming, or deflecting responsibilities of important kinds, and understanding the implications of doing so" (Urban Walker, 1996, 276). Of course, this idea is contrasted with the so-called "theoretical-juridical conception of morality."

[10] The role of ethics of care in bioethics is illustrated in Carse, 1999.

beings in a similarly conscientious fashion. Consequently, even a lowly kitchen vase can be considered a moral mediator in the sense I give to this cognitive concept.

When I clean my computer, I am caring for it because of its economical worth and its value as a tool for other humans. When, on the other hand, I use my computer as an epistemic or cognitive mediator for my research or didactic activities, I am considering its intellectual prosthetic worth.

To make a case for respecting people as we respect computers, we can call attention to the values human beings have in common with these machines:

1) humans beings are – biological – "tools" with economic and instrumental value, and as such, can be "used" to teach and inform others much the way we use hardware and software, so humans are instrumentally precious for other humans in sharing skills of various kinds; and

2) like computers, people are skilful problem solvers imbued with the moral and intrinsic worth of cognition.

7. Conclusion

The main thesis of this paper is that in recent times, non-human beings, objects, and structures like technological artefacts and machines have acquired new moral worth and intrinsic values. Kantian tradition in ethics teaches that human beings do not have to be treated solely as "means", or as "things", that is, in a merely instrumental way, but also have to be treated as "ends". I contend that human beings can be treated as "things" in the sense that they have to be "respected" as things are sometimes. People have to reclaim instrumental and moral values already enjoyed by external things and objects. This is central to the aim of reconfiguring human dignity in our technological world. Aiming at illustrating the intrigue of this ethical struggle between human beings and things I have discussed the role of objects, structures, and technological artefacts by presenting them as *moral carriers* and *mediators*. I maintain this perspective can be very fruitful to approach many other problems related to the relationships between machines and ethics

A PHILOSOPHER'S TAKE ON MACHINE CONSCIOUSNESS[1]

PETER BOLTUC

1. The Two Senses of Consciousness

There are two different, though not incompatible, ways in which someone can be conscious.

In the first, most commonly used, sense one is conscious if one performs certain cognitive functions. Thresholds and descriptions of what counts as such functions differ since we encounter somewhat different distinctions between consciousness and cognition in the literature of the subject [Boltuc/Boltuc 2007]. In general one expects the capacity for learning, hence memory, and other more or less complex feedback loops that distinguish conscious from merely cognitive structures.

This first sense of consciousness is where the focus of research in machine consciousness lies, and rightly so. In AI at least one of the focal points is to replicate cognitive processes known from human thinking and then even to improve on them, for instance in terms of speed and accuracy [Baars][2].

The second sense of consciousness pertains to consciousness as the first-person stream of awareness. This sense has been barely present in discussions of machine consciousness, except in a dismissively unhelpful manner. While this sense is hard to overlook in common-sense everyday intuitions, it is also hard to clearly present within the Popperian, or any verificationist, model of science. Therefore, it has been dismissed by many

---

[1] The paper was presented at PHAEDE (Suceava, 02, 2009). Jagiellonian University (Krakow, 05, 2009), and NA-CAP at the APA (Chicago, 02, 2010). I am thankful to all participants and especially J. Barker, J. Bremer, M. Kuniński, T. Polger, R. Sanz and R. Turner for their comments and suggestions.

[2] The literature on this subject is copious. In the current paper I provide only the literature I directly refer to while a more robust literature review can be found in [Boltuc 2009].

philosophers and psychologists, although more readily so in the mid 20th than in the early 21st century.

The main bulk of this article is devoted to the defence of the naturalistic version of consciousness as first-person awareness.

2. Philosophical Issues Pertaining to the First Sense of Consciousness

Philosophically the first, functionalist, approach to consciousness is interesting in many ways, but two programs seem most commonly shared:

Firstly, it is important for philosophers to dispel bad philosophical ideas that pock holes or even try to stifle the worthy project of constructing artificial intelligence on some half-baked philosophical grounds. Already Turing tried to deal with hole-pocking objections, such as the one that machines could never perform some cognitive operations ('think') as well as humans do [Turing].

Such objections are embarrassingly outdated now, and philosophers are less and less needed to demonstrate this. Developments in AI and other areas of artificial consciousness provide the best answer to all those hole-pocking attempts: that such cognitive limits on artificial minds do not exist.[3] The proof is in the pudding since the newer and newer tests of what machines can't do are being met by programs performing such varying tasks as driving vehicles, fixing one's clogged arteries, providing daily companionship or writing classical music [Floridi]. Hence, endeavours centred around the hole-pocking project – trying to defend the final frontier of what machines certainly could not do – may provide interesting engineering challenges but no longer do they hold any direct philosophical relevance.

Secondly, there is some needed meta-reflection on this kind of machine consciousness; it is mostly axiological reflection. While moral evaluation of various aspects of machine consciousness, for instance in terms of acceptable risks, is vital for incorporating robotics and machine consciousness in general into the web of interactions of human beings with the world [Johnson and Miller; Moore], sometimes evaluators put themselves in the position of judge and jury and give insufficient attention to the potentially harmful effects of their recommendations to the discipline of machine consciousness. Much of the potential stifling effect on new research in bioengineering and related fields brought about by philosophy

[3] In [Boltuc 2010b] I show why the fact that no machine has quite exactly met the Turing Test yet provides no real counterexample to this claim.

comes from ethical and social considerations[4]. The stifling effect is felt in many engineering fields: it has brought the whole area of human genetic improvements nearly to the standstill. Similarly, some philosophers still worry what conscious machines would do to human primacy in the universe; a question which I find incredibly unimportant.

There is also a third, less broadly shared, sense in which philosophical and psychological understanding of consciousness is important for machine consciousness. As [Sanz] recently pointed out, advanced engineering of conscious machines needs to learn much from philosophy of mind. The main reason why machines may still be unable to perform complex autonomous tasks such as driving a car, or military operations in a complex environment, is not so much a function of complexity of those tasks. It rather comes from the fact that 'they do not know what is going on' – that they lack a broader context. A rather dumb human can drive a car just fine whereas a very sophisticated robot can do so only on desert highways, away from real intricacies of city-traffic. The problem is that robots do not understand enough of the context (what a little old lady trying to cross the road, a person who drops her bag on the sidewalk, or kids playing ball nearby, may do and how this could interfere not just with one's own driving path but also with the driving of others on the road). In order to function in the human world, where all interactions are partly social interactions with socio-cultural contexts, one (including a robot) must be a part of the network of such interactions, or, at the very least must have a working understanding of the complex, somewhat informal, 'language' of socio-cultural networks of events as well as somewhat intricate psycho-logical features of other players involved. Such knowledge, however simple it may seem to us, takes a complex cognitive system to acquire and use properly [Sanz].

The direction of fit between the third, and the two previously mentioned uses of philosophy is different. In the first two instances philosophy provided meta-reflection on machine consciousness, either in terms of hole-pocking (which may be less fruitful now, but helped provide a set of solid answers to the question of whether machines could in principle generate all kinds of cognitive operations that human beings perform), or in terms of ethical reflection on machine consciousness and

[4] Some of those resemble the recent Swiss law requiring that anybody causing changes in ecosystems needs to prove in court that ecosystems, or their members, for instance trees, would want proposed changes happen to them; this I believe is a misguided approach and similar approaches to machine consciousness would also be quite misguided; as long as machines are not h-conscious.

its applications. In the third sense it is consciousness-engineering that returns to the philosophical, psychological and neuro-scientific theory of consciousness so as to learn how the most advanced forms of consciousness that we know of, which are still biological, deal with all sorts of cognitive tasks provided by the world [Haikonen].

Those three issues are not supposed to cover the field of philosophy of computer science, or even of machine consciousness. There are better known philosophical issues in computer science, such as the halting problem, or whether there could be exceptions to the Church-Turing thesis (whether every information processing is Turing-computable, or whether some non-standard mathematical procedures aren't). But their practical relevance for machine consciousness occurs through mathematics or logic that underlies the engineering level. The issues of machine consciousness listed above belong more directly to philosophy of mind. Yet they are all 'functional' issues, related to the way consciousness operates – they refer to the first sense of consciousness.

3. The Philosophical Issue Pertaining to the Second Sense of Consciousness

It seems that engineering (AI or otherwise) is far from being able to build a machine equipped with the sort of consciousness covered by the second definition, at least not any time soon, except maybe for cyborgs. This is in part because nobody seems to be trying to undertake such a project, but largely because we seem quite far away from even understanding how exactly such first-person consciousness works in animals. Theories of how consciousness operates and where it is generated abound but none has been commonly accepted, except for generalities such as that it is largely associated with the thalamus.

Here is the philosophical backdrop of this paper: *the dismissively unhelpful* eliminativist attitude towards first-person consciousness adopted, mostly due to narrow verificationism, by behaviourists and some functionalists in philosophy of mind, makes it particularly important to deal with this issue anew. A number of people both in AI and philosophy seem to conflate the very well grounded claim that computers can in principle conduct all sorts of thought operations that humans can in principle do (they rightly reject the hole-pocking approach), with the claim that the current sort of computers or robots are (or, by improving on certain tasks, could become) conscious, and in the same way humans and certain other animals are at that!

A more sophisticated version of the functionalist approach includes phenomenal consciousness. It maintains that robots, which are equipped with a camera or other sensors allowing them to be guided by sense-data such as colour, smell or temperature, have the first-person consciousness. Unfortunately, most uses of this view confuse the two senses of the term 'the first-person perspective.' In the first sense, the first person point of view is just a position at which a camera or other type of sensors can be located; they present an image of the world from a given place, with specific coordinates, such as the view of a certain mountain from point A one mile just west of it, as opposed to its views from all other possible viewpoints. Let us call this *the positional sense of first person consciousness*. It is an application of the first sense of consciousness discusses in section 1 of this paper.

In the second sense, the first person point of view is linked to awareness. It is a set of experiences, such as auditory, tactile and visual experiences that appear in the state of being aware or conscious in humans and many animals. Let us call it *h-consciousness*, or *first person awareness*. It is an application of the second sense of consciousness discussed in section 1 of this paper.

The reason for the name h-consciousness is complicated and thoroughly explained elsewhere [Boltuc 2009]. In the current paper I want to avoid as much as possible the conversation about various senses of phenomenal consciousness [Block], since the notion of phenomenal consciousness also has both, the interpretations that are purely functional (the strictly positional sense), and those that pertain to the special feel of first-person aware experiences. The reason for calling the latter sense h-consciousness is involved with philosophical debates; namely, with the well-known argument by D. Chalmers that distinguishes between the hard and the easy problem of consciousness. I do not subscribe to all nukes and crannies of Chalmers' argument[5] (by now, he has abandoned some of them as well) but I do think the notion of consciousness he points out to, and the intuitions he provides are very strong. The other strong intuitions come from T. Nagel but his older work may still be viewed as confusing the two senses of first-person perspective distinguished above. Hence, in 2007

[5] [Dennett] demonstrated that Chalmers' use of the case of zombies begs the question against materialism (since it violates the tenet of materialism that there is no difference without physical difference). [Sloman] shows that the 'easy problem' of consciousness is not easy (with which Chalmers agrees) and that the claim that the 'hard problem' is conceptually at a different level is somewhat presumptuous, though I think Sloman's conclusion that therefore 'the hard problem' either does not exist or is at the same level is presumptuous as well.

[Boltuc and Boltuc] we introduced the notion of h-consciousness to indicate the new, not just positional, use of this term similar to Chalmers' notion defended in his arguments for the hard problem of consciousness, yet without endorsing all of the intricacies of those arguments.

I show elsewhere [Boltuc 2009; Boltuc 2008] that functional arguments for h-consciousness are necessarily deficient; this is because h-consciousness is non-functional, it is not definable by any third-person observable function. This includes the so-called Black and White Mary Case, the zombies case and all other instances where one searches for functional deficiencies that only the notion of first-person awareness could resolve. I argue that while such deficiencies may indeed be guided by first-person awareness (since they emerge naturally in the order of discovery guided by its experiences), those deficiencies as well as the solutions provided by first personal awareness can all be replicated (faked, if one wants to be presumptuous) by functionally identical systems without awareness. My main difference with Chalmers is that he seems to still believe in the project of finding functional differences between the two perspectives which would distinguish h-consciousness from the third-person perspective.

One may ask, naturally, whether h-consciousness is just epiphenomenal[6]. This question deserves more of a response than I am able to provide here, but here is a draft: Clearly, all possible functions of h-consciousness could be replicated in the purely functional consciousness equipped with the positional first person consciousness yet devoid of awareness. Yet, the driving force of why to choose any given notion comes from aware experiences.

This answer may seem unsatisfactory as long as we presume that there exist a small number of such options so that what counts as the 'driving force' is of low importance since all options would have been discovered by purely functional structures. The situation looks different if the ontology of possible eventualities is very large and the choice of which ones to instantiate in the world is very non-trivial. In such systems what counts as the driving force does make an important difference; the claim is that the first person aware perspective provides subjects with the insights what choices to make, which would be hard to come by according to strictly third person criteria. First person consciousness figures out guiding potential choices, though such choices may be graphed *ex-post* by some

---

[6] I owe this question to J. Bremer and discussion of some parts of the answer to T. Polger..

complex, maybe non-linear function; yet they can hardly be extrapolated by such function.

Consciousness as awareness is important also in another closely related way: it provides second order inductive evidence. Due to the privileged access to the first-person perspective, one has *direct* access only to one's own stream of awareness. However, as psychology understands better and better, we can measure such consciousness based on second-order research that compares testimonies of various subjects. The fact that such testimonies could be faked does not lead to the denial that good inductive evidence for such awareness exists, can be further collected, and verified within, always fallible, inductive criteria of confirmation.

Concluding this section, in order to appreciate its practical importance, the above point needs to be viewed against the backdrop of the following claims. The more or less reductivist approaches, invited by behaviourism, most forms of functionalism and other well-known currents or undercurrents of the 20th century philosophy of mind, create a clear danger of excessive (even if non-eliminativist) reductionism in philosophy of mind[7]. We are faced with a resulting vision that there is nothing to human consciousness that the more sophisticated of present day robots (or those to be expected soon) do not possess. This claim has far reaching implications for the (mis)understanding of human consciousness, survival and ethical concerns that seem largely, if indirectly, related to the status of a subject as a consciousness [Boltuc 2008].

4. Introduction to H-consciousness.

A. Historical Inspirations

In this section we try to further specify this second sense of consciousness, the one missing in the philosophical debates informed by functionalism and behaviourism, if not eliminativism. Fichte and Husserl inspired my answer to this question, though what follows should not be viewed as a piece of history of philosophy, but rather as a humble attempt to follow up on the philosophical intuitions, which owes much inspiration to their work. Those intuitions seem more up to date than Descartes' substance dualism, still talked about, and, even by Chalmers, taken with

[7] Which reductionism provides an additional reason to view these approaches with suspicion.

all too much epistemological respect[8]. Instead, the view of 'pure subject' shared, in different versions, by Fichte and Husserl (and a few other, mostly German, philosophers) seems to respond to the philosophical need to express the gist of the first person perspective, and to do so to the greater extent than anything available to analytical philosophers[9].

What makes the view of Fichte (present, for instance, in his *Second Introduction to the Theory of Knowledge*) and Husserl (mostly in his *Ideas*) so relevant here is their (strikingly similar in its very gist) intuition expressed in the notion of pure epistemic subject. The point is that we should look for the simplest or most reduced notion of epistemic subjectivity: the subject that is, as nearly as possible, not an object at all.

This is the search for the most reduced or basic notion of the subject we can attain. While Kant may have a similar notion to Fichte, I am persuaded by Siemek's argument that the latter's notion of pure transcendental subject is more reduced. Kant's active subject is of course a much more intricate, higher order, notion due to its function in co-constitution of the reality. I am open to the argument that Kant has also the notion of pure epistemic subject similar to that of Fichte and Husserl, and that the notion is already present in Kant's First Critique (I would actually be quite happy if this was the case). Yet, the point of this paper is emphatically not to make any inroads into the history of philosophy and to remain merely, but programmatically, at the level of broad, even if vague or historically controversial, philosophical inspirations.

B. The four ways to support h-consciousness

I have four ways to support first person awareness (or pure epistemic subject), and maybe even arguments for it, which I partly developed and largely compiled and collected over the years.

1. The first argument is of a negative kind; it was mentioned above but I return to it here for the sake of clarity. I think all the attempts to show a functional need for first person consciousness are not quite satisfactory due to a broader theoretical reason, though they may serve as helpful (though somewhat misleading) intuition pumps. The, now largely defunct,

---

[8] Substance dualism may still be a respectable view in philosophy of religion, or religious exegesis, the way Swinburne or Taliaferro approach it, but not quite as a viable option in epistemology.

[9] Despite fruitful attempts for instance by Strawson, Nagel (mostly in *The View from Nowhere* and the paper "Subjective and Objective", not the notorious "What is It Like to be a Bat?" paper) and a few others.

Black and White Mary's case or the case of zombies suffer from the same deficiency – they try to search for functional deficiencies left by first person awareness (in this case viewed as non-reducible *qualia*). I used to partake in the illusion that such attempts could possible succeed [Boltuc 1998a, 1998b, 1987]. Yet, the assumption that first-person perspective leaves anything open in the functional world is misleading, if the subject is not an object at all [Boltuc 2009, 2008] and this is what the Fichtean-Husselrian intuition amounts to.

As I mention before, this view may lead to strict epiphenomenalism about the subject, but let me revisit this issue as well. I think there is a different option here: The most plausible interpretation is that first-person awareness carries the true causal role in causing various states, for instance those evoked in the Black and White Mary's case when having seen her first red tomato she does acquire a skill in recognizing red objects by visual inspection. Yet, the view that there is a functional deficiency that can be covered *solely* by the first-person awareness hypothesis is misleading. The functional description, though not causally efficient in the normal instances, can always be built upon (faked, if you may) any functional deficiencies [Boltuc 2010a].

2. The second argument may be called ontology of the epistemic subject. I define the epistemic subject as such a subject that is not an object of the sort that we can directly predicate about. I describe various ontological reference frames ('universes') in which such a subject can be indirectly defined. Then I search for the simplest such reference frame. This is to show what role the metaphysical standing of the epistemic subject plays, indirectly, in the ontological framework, since some authors seems to view it as merely epistemic and illusory in that way. I understand that those remarks remain somewhat opaque, but I must relate the reader to a longer argument elsewhere [Boltuc 2009]. I also think that there are somewhat broadly shared intuitions about the role of such subject as the locus of non-instrumental value [Boltuc 2008].

3. The third argument may be called the search for the pure epistemic subject. It is an epistemic counterpart to the second argument (searching for the ontological approach to the epistemic subject). It starts with phenomenology similar to Descartes' but avoids jumping into ontology of any, and particularly of the dualist, kind.

4. The fourth argument operates within philosophical anthropology; we focus on the various notions of the subject[10] and look for the simplest one available. Hence, we look for a subject that is not an object but not through ontological reference frames but by distilling it through various more robust notions of the subject.

In this paper I focus supporting the third and fourth argument; the first two arguments I developed elsewhere [Boltuc 2009, 2008].

5. The Epistemic (Even Phenomenalist) Argument for H-consciousness

This is the most travelled route towards non-reductionism, and perhaps the most mishandled argument. The problem is that the arguments for non-reductive consciousness based on first-person insights are taken to mean way too much in a broader context. The authors who have strong phenomenalist intuitions, perhaps except for Russell, tend to have a hard time viewing the first person epistemic standpoint only as a piece of evidence in a broader inductive framework. My point here is to sanitize them from superfluous, misleading over-interpretations. Hence, the epistemic or phenomenalist argument is only able to provide important supporting evidence for a broader, inductive, case that can be made for non-reductionism.

Here is a 'thought experiment' often given to undergraduate students. Try not to think about anything in particular. If you can attain this state of mind (some people can't but others, mostly those who did some Buddhist training, transcendental meditation or just relaxation techniques, are able to), you can have moments of consciousness without content. This is the closest we can come to gaining the intuitive grasp of pure consciousness, or a subject without an object. [11] This is nothing more than an intuition pump since pure subject is ontologically impossible without an object (Kant, Fichte, Hartmann and others would say, that also an object is impossible without a subject but this point goes beyond the scope of this article). Yet, this intuition pump helps us grasp the subjective side of the epistemic relation. Many philosophers cry out loud that such analysis leads to dualism, but they are trusting Descartes way too much.

The primary subject-object relationship puts subject and object as the two necessary components of any experience. Deep ontological

[10] An early version of this argument is in my first Ph.D. dissertation [Boltuc 1991].

[11] I remember this point, made persuasively by a Soviet philosopher Misha Konkin at a conference in 1987.

implications, certainly those extending all the way towards dualism, do not follow. The subject and object are nothing more than the analytical aspects of experience.

The purely epistemic point of view does not allow us to reach beyond the experience, and all the objects revealed by it are, in the final account, always only the objects of experience. The ontological status of such objects, as those outside of experience, is ill defined; some authors even claim[12] that the notion of the external world can't be formulated in purely epistemic frameworks. I would rather argue, leaving aside causal theory of reference, that ontologies outside of the epistemic framework can be formulated but always as a more or less supported inductive hypothesis. The epistemic reference frame does not need to fall into scepticism about the external world but it will always view the latter as a product of an inductive hypothesis.

Suppose that you think you see some flowers in the room. They can be flowers, but they can be hallucinations induced in one of the many (chemical, mechanical, electro-magnetic and so on) ways directly into one's brain, or they may be physical objects of a completely different kind, such as a hologram. Within the epistemic, purely phenomenalist, perspective it is always an open question which of those and infinitely many even less supported hypotheses happens to be the case; only overwhelming evidence gives us inductive (always defeasible) reasons to believe what their status is, which is never anything close to metaphysical certainty.

This is a flip side of the problem of privileged access encountered from the ontological perspective. Just like from the epistemic viewpoint we can never 'really know' (beyond inductive evidence) that the external world exists, from the ontological point of view we can never really know that any subject has his/her first person awareness; but also in the latter case we can have good inductive evidence, based on behaviours, self-reports and physiological evidence, to this effect.

There is an analogy between the epistemic perspective's inability to really justify (following strong metaphysical standards) the existence of the external world and the ontological perspective's inability to justify the first person epistemic awareness[13]. Hence, from the epistemic viewpoint pure subject is the locus of the epistemic perspective, or the locus of

---

[12] So does H. Putnam, in dealing with a somewhat analogues question raised in his *brain in a vat* example.

[13] Russell's neutral monism is an ingenious attempt to build a complementary perspective between those two views, which I follow, loosely, in my argument.

awareness. It can be viewed, imperfectly but intuitively, as the consciousness in as much as possible devoid of the objects of consciousness.

Now, what is the relation between this claim and the argument made, so famously, by Descartes? There are two points to Descartes' argument: the epistemic and the ontological point. Many philosophers believe that the epistemic non-reductionism put forth in his examples leads to epistemic dualism, but this belief is far from the truth. The assumption, that if there is an epistemic subject it must be a different 'substance' in some mysterious sense is a *non sequitur.* The dualism of substance is just one, and, as we know now, pretty crude, interpretation of epistemic non-reductionism. In this paper I argue for a naturalist vision of epistemic non-reductionism.

6. The Levels of the Subject

We can now develop the fourth approach, which helps us clarify the most reduced notion of 'pure' subject. The argument is based on the comparison of various definitions of the subject; it distinguishes it from other related notions. Some authors tend to mix those notions and equivocate among them. Let us try and sort them out[14].

A. Social subject and the social definition of a subject

The meaning of the word subject reaches the furthest away from the meaning of pure epistemic subject when we talk about the social subject; for instance, Marxists talk about social classes, such as the proletariat, as subjects of social processes. Here a subject is a group with certain cohesive features. Such subjects, discussed in sociology but also social philosophy, may in fact be subjects in a broad sense of the word, for instance they may be sources of action. The problem arises if authors conflate this meaning of the subject with those presented below.

There is also another meaning of the social definition of a subject. In this second sense an individual human being can be defined through the social context (this is the so-called *dossier-man*, a person defined by her dossier or business card). Here we define a person exclusively through social relations, or the social role, she holds.

---

[14] I make this point in my 1991 dissertation (in Polish) and in the [Boltuc 1990] article.

B. Physical subjects – physicalism *versus* naturalism

Physical subjects are defined as objects in the world. Analogously to the social definition of the subject, also physical subjects can be constituted by groups, for instance a certain group of individuals, defined in their physical sense, may be viewed as the source of some process, for instance of putting together a pile of stones. Yet, the main sense I which the notion of physical subject is used in philosophy of mind is, in the context of physicalism, as an individual physical being. Physicalism is, largely, a claim that all features of a person are the features of her physical body.

Whatever one's view on physicalism one must accept the view that at one level of description human beings, and other persons we know, can be perceived and described as physical bodies. Moreover, achievements of neuroscience, medicine and other disciplines dealing largely with human bodies indicate that this is an important level of description.

The difference between physicalism and naturalism is hard to grasp but important. Usually, physicalism is viewed as reductive materialism, the view that all features are somehow reducible to the physical features. For instance, it is a stretch to talk about physicalist emergentism since the latter includes the properties not directly reducible to physical properties. On the other hand, there is no analogous problem with naturalistic emergentism understood for instance as the view that nature works in such way that sometimes certain sets of properties emerge from lower level properties in a non-linear fashion[15].

Emergentism is just an example of a relevant difference between naturalism and physicalism though the distinction still suffers from much vagueness. There seems to be also a methodological difference: The main tenet of physicalist materialism in a broader sense is that there is no difference without physical difference; naturalism is likely to adopt this claim but not as an assumption, but as a fact based on collected evidence.

C. Psychological subject and the brain view

The next, more reduced, level of the subject pertains to one's psychological features. A subject (sometimes viewed as a person) is identified with one's psychological features or states. There are of course different versions of this view that follow different positions on human

[15] I owe this point to a spirited discussion among a number of participants of the 2009 Naturalism Workshop in Kazimierz.

psychology: some focus on emotions (Gilligan), some on language capabilities (Dawidson) or intellect (Kohlberg). Some of those approaches view human psychology as a direct product of one's cognitive apparatus (Haidt) and a function of it. While some psychological views on the subject presume relative autonomy of human *psyche*, most are close to reductive materialism. What makes the latter approach different than physicalism is rather the primary focus on psychological features than any theoretical disagreements with physicalist methodology.

Here we do not need to plunge into these debates. It suffices to say that a subject, or a consciousness, can be viewed primarily at its psychological level, whatever the background ontologies of such views may be.

It is sometimes argued, primarily on the basis of the psychological criterion, that a person is primarily her brain (or central nervous system). This is because brain or CNS instantiates psychological features that are, philosophically speaking, viewed as central to a person.

D. The stream of consciousness

A step more reduced view than the psychological criterion of personal identity is the view that we are really the stream of our conscious experiences. To apply this position to the task of defining consciousness, in accordance to this view one's consciousness is one's stream of consciousness. Famously, Hume, and today Parfit, make interesting arguments for this view.

According to Parfit one is defined by continuity and connectedness of one's experiences (and moreover, identity of such stream of consciousness is not really what matters, or is attainable, since branching is possible). This is a follow up on earlier versions of phenomenalism that reduce one's consciousness to the stream of conscious experiences. The counterargument to this approach has been provided by the so-called case of the experience machine: If one could have all the best experience during one's lifetime but only induced into his/her brain by a machine, with no correlates in the outside world, would it be a good life[16]? While some people say yes, most believe that purely epistemic experiences need to have the ontological counterpart to count as fully satisfactory – they must be true experiences by some version of Tarski's definition of the truth. The experiences matter in part as indicators of the ontological state of affairs, not just on the basis of the content of the stream of consciousness. (This was the problem with uses of the epistemic, phenomenalist, arguments for pure consciousness

[16] The issue was taken up by the Wachowski brothers in *The Matrix*.

discussed in section 5 if taken aside from the other arguments, and in particular from some form of ontology.)

E. Transcendental active subject

According to Kant the subject co-constitutes the world by putting the experiences into the categories such as time, space and causality. By now we know, from Einstein, that time and space are not merely the categories of apperception: they are physical categories and non-absolute at that. Yet, it makes sense that the mind imposes some of its perceptual framework onto the input it receives from the world, a bit like a computer deals with the impulses acquired from the net.

Hence, the category of active subject makes sense even if one does not buy into the whole metaphysical picture of Kant's transcendentalism. However, it is important to notice that the transcendental subject is not the most reduced notion of the subject, in part because it must be structured complex enough to perform the act of shaping the world as we know it into the categories of time, space, causality or even justice.

F. The mixed approach to personal identity

This is an argument on various definitions of consciousness or various levels of generality. It does not aim at establishing the one 'true' or 'correct' level at which the right definition of consciousness lies[17]. To the contrary, all those levels describe a certain valuable level. The point is just not to cut off any of those levels, at last not without the sufficient reasons. This pertains, emphatically, to the level of pure subject to which we now proceed.

G. Pure subject

The most reduced category of the subject is the subject that is solely the epistemic pole of the basic epistemic relation or the subject-object relationship. Pure subject is not an object at all. We can know of it only by inference [Boltuc 2009, 2008, 1987].

---

[17] The problem is closely related to the issue of personal identity. Also in the latter problem the mixed definitions providing some physical and some psychological criteria seem to fare better in a long run (except on the account of simplicity) than the neat monistic views.

Such notion does lead to non-reductionism, in fact it is the gist of it. Yet, it does not entail any broader versions of metaphysical dualism. Subject is not a substance but rather a necessary condition of the possibility of every epistemic process, or awareness. Without assuming it we beg the question in favour of reductionism. The reason we have for adopting it is, largely, our first-person experience of being aware. It makes sense to assume that certain beings are aware and others are not

I tend to call the notion of pure subject H-consciousness, which is to situate it in the current debate, mostly between Chalmers and Block. In some ways a clearer term is awareness or first-person consciousness but those terms are acquired throuth some strictly functional definition, in philosophy or psychology. Hence, in order to express the strictly first-personal character of this notion we need a technical term: h-consciousness fits the bill.

7. Defining the engineering thesis for H-consciousness

There should be a reason to assume that certain beings, such as humans and higher animals, have first person consciousness, while others, such as door knobs, do not. While Chalmers experiments with a panpsychic view that all objects have some consciousness, though human and animal brains are very good in gaining more of it than most other objects do, the arguments in favour of this approach are a bit skimpy [Chalmers]. It is a better supported inductive hypothesis that there is a mechanism in animal brains that is somewhat unique in creating first person awareness. While we do not presently know how exactly this mechanism operates, we know enough about it to have good reasons to believe that such knowledge will be gained by science eventually. Moreover, the past developments in science demonstrate that the kinds of question of how various things, or functions of animals, operate sooner or later gain precise scientific answers.

Does this mean that I agree with Sloman and others who claim that there is no difference between Chalmers' hard and easy problem of consciousness? Not quite; the truth in Chalmers' approach is his stand in favour of a non-reductive view on consciousness. Hence, Chalmers is right that it is impossible to explain out the non-reductive features of first person consciousness. The problem is that he seems to search for some such explanation, whereas I opt for the non-reductive but extremely narrow notion of first-person consciousness, for the notion of pure epistemic subject. I am not sure that there is more to the hard problem of consciousness than to accept this Fichtean-Husserlian claim.

What can be explained is the mechanism in which the first person consciousness is produced. Such explanation belongs to the level of the easy problem of consciousness, to use Chalmers' phrase, though it does not pertain to the project of finding neural correlates of consciousness, which Chalmers understands as specific conscious states. We are now talking of the neural correlate of consciousness *tout court*, namely, of the correlate of awareness, or the mechanism producing it.

The engineering thesis in machine consciousness [Boltuc 2010, 2009; Boltuc/Boltuc 2007] is the claim that once we can understand how first person awareness is produced we should be able to engineer it. It is a technical question when this would happen, or whether it can be engineered in inorganic, or only in organic matter. Hence, we need to acknowledge the following:

1. There is no reason for naturalists to dismiss non-reductive views on consciousness and many reasons to adopt it.
2. There are very strong reasons to believe that first person consciousness is a natural process.
3. If 2, then we should be able to get to understand it in the future.
4. If 2 and 3 then we should be able to engineer it in a somewhat distant future.

The reason why this point is important now is that many philosophers and engineers seem to deny 1, which leads to endorsing reductive claims on consciousness. It also leads to the attempts to present purely functional consciousnesses (functionalities) of some computers, robots or electronic agents as conscious. While those activities may be impressive in terms of the tasks such agents perform, and while they satisfy the functional definition of consciousness, we should be warned against claiming that by this fact they satisfy the definition of consciousness tout court.

It is a possible long-term program to produce machines with first person awareness but there is absolutely no reason to believe that any contemporary robots or computers even come close to attaining this, since their construction does not involve the thalamus-like structures that would give them a chance for such function as first-person consciousness. It is a different topic entirely whether such a program would make sense to perform, or be morally all right to do so. I tend to pose a tentative yes on both accounts. First, since I claim that first person awareness does play an indirect functional role, by picking the options from among the vast set of possibilities, which purely functional robots may later imitate, I claim that involving machine awareness may produce new functionalities. The pure awareness of course would not do the job, but a system, a first person psychology, equipped with such awareness would likely provide a machine with new patterns helpful in selecting from among the nearly infinite possibilities of action. Second, while some ethical constraints, and safety constraints, are justified, we should take into account two facts. The moral value attributable to a first-person consciousness is limited since most of us attribute very little inherent moral value to animals such as rats that doubtlessly possess first-person consciousness. Second, as I mention above, the stifling effect of overzealous ethicists probably brings much more harm (wasted opportunities) than benefits.

COGNITIVE LIFE RE-ENGINEERED

COLIN T. A. SCHMIDT

Preamble on Technology and Change

European societies have been at the forefront of industrial-like technological revolutions. They have the wherewithal to make and break movements in this area and they have a tendency to produce knowledge navigators and originality in this area. It would seem that technical objects were difficult to deal with in other societies, heavy to accommodate and sometimes poorly accepted by the people at the helm of various scientific activities. This of course did not render it easy to fathom newer facets of technology; especially anything that could be seen as granting freedom or labelled as autonomous. Many things have changed in Western Europe since then. One could say that change of such importance was less necessary in Eastern Europe for highlighting the principles and phenomena increasingly evident to us in cognitive studies, those that we are about to work with here. The text that follows looks into aspects of communication within our society along with some difficulties that could occur due to newer aspects of our lives, manners and know-how.

The focus on the use of language has been chosen because it is the most demonstrative element of cognition; verbal activity is expressive and lends itself nicely to writing about intelligence.

The Role of Autonomous Technology in Society

Historically, society did not perceive or acknowledge the value of technical objects; it was not until early in the twentieth century that the feeling society had for such products started to turn positive.

The reason for the lateness of this attention given to technical objects *latu senso* resides in the semantics they carry, especially in European societies:

1. hot, dirty and often infested factories;
2. sweat and hard labour;
3. infernal cadences imposed by Tayloristic theories of the workplace;

4. the image of members of the lower classes that traditionally work in these contexts; and so on...

Of the listed negative meanings that technical products convey, the first two components are basically due to the nature of the final product – heavy objects, machinery and of course the quantity of brute material necessary to build them. The fourth point is blurring out (or is gone, as in the New World). The third negative component mentioned has not totally vanished, neo-Taylorists – or other suitable labels for the 'rigorous bosses' of modern industry – respond as present in today's society. The difference then, when one looks at the evolution of technical objects, take for example evocative ones like simple automata one hundred years ago and current-day machinery that have the gift of speech, learning, "reasoning", etc., is that a lot of tasks have been automated and miniaturised. The image of the person working day and night to design and build a machine that will have all the human traits necessary for providing company to lonely persons is much more bearable and somewhat glamorous. Many fields necessitating artificial neural systems have recently sprung up in this area (i.e. "service robotics", "autonomous agents" ...).

The age being modelled over time for discourse with old and new human-like 'contraptions' is as followed:

1. *1950s* A. Turing • Adult Discourse
2. *1960s -70s* J. Weizenbaum • Adult/Teenager Discourse?
3. *1980s* various authors • 7 to 15 yrs. olds
4. *1990s* R. Brooks team • 4-7 yrs. olds
5. *2000s* J. Zlatev • 2 yrs. olds
6. *2001* B. Scassellati • > 1 yr. old
7. *2007->* ? • > ? months old

It seems there is some regression here as scientists modify their expectations of what the machine can do.

So although technical objects as we know them did get a weak start, with time we have learned to procure a more 'elegant' place for them in our society and in the scientific community. In fact, this latter may have got carried away with attributing more or less social roles to artificially neural-rich machines. Does this mean they are gaining autonomy thanks to Man? Is it possible for a technical object to gain autonomy, as humans, if it were made by a human being? To what extent is the use of the word "autonomous" meant to be simply metaphorical[1]?

Some questions of the philosophical sort are starting to be raised in Cognitive Science, cognitive robotics, etc. In this sense, one could say that the year 2001 was a "fast year" for research in Robotics. In that year, J. Zlatev, author of an article in *Minds and Machines,* asks two highly pertinent questions for robotics:

1. If a robot is able to participate in simple language games as adequately as a child, should we concede that the robot handles true meaning? and
2. How would we go about developing a robot which could possibly live up to a positive answer to the first question?

My approach is straightforward: *a)* refute the first question, so as to *b)* be able to drop the last. This rhetorical statement is meant to draw attention to a growing problem due to computation-related communities that valorise materialism beyond necessity. As the author of the questions above inspires the Epigenetic Robotics movement – movement for which it is preferable that a robot *learn* the appropriate behaviour rather than simply having it *programmed in* –, it would seem that what he meant by saying "developing a robot (able to participate in language games)" is that learning is *the* most important aspect when it comes to integrating intelligence "into matter".

Another author named P.M. Milner, perhaps better known to the readership of *Neuro-computing*, expresses the limits of betting heavily on the learning explanation horse:

> We should recognise that learning is a relatively recent evolutionary development, and that most of the animal population, including some of the most successful species, flourish with negligible capacity for individual learning. Some build snares, for example, or communal dwellings that would tax the ingenuity and skill of a human [...] Learning is not a substitute for innate behaviour; it is an example of it. Learning is an evolutionary development that allows fine tuning of a very complex piece of predominantly heritable machinery. (Milner 1999, 6-7)

It is possible that Computer Scientists and roboticists are over-enthusiastic about learning as it is even more recent in their discipline.

---

[1] I treat these 'loaded' questions elsewhere with F. Kraemer, (Schmidt, C.T.A. and Kraemer, F. 2006). *Cf.* also: Schmidt, C.T. 2004.

I myself think the problem bringing up such a debate for computing, robotics, etc. resides in the fact that a large part of intelligence cannot be situated in the matter of the individual, particularly the dialogical components of it. I therefore argue in favour of supporting another well-known sub-domain of AI/HCI/Robotics thought in order to stimulate research in the artificial sciences based on the reality the discursive aspects of the human mind bring forth (*cf. infra* Preamble). This reality should have technological consequences. I am sure the reader will agree with me that, after reading this article, some of the actions produced in recently-formed scientific communities could be rethought for the future.

Robotic Brains under the Projectors

I would now like to speak about an issue that has a 50-year and more history in the Sciences of the Artificial. Important research being carried out at top-notch scientific institutions like MIT, Carnegie Mellon University and still yet many others seem to be having difficulty with the mind-body problem in creating robots that think. Weng, McClelland, Pentland, Sporns, Stockman, Sur and Thelen teamed up to confirm this in their *Science Magazine* article a few years back (2001) with discussion on "autonomous mental development" that was limited to brain and body building (Weng, J., et al. 2001, 599-600). Whether their intention included outright occultation of the *mind* or not, *reductionism* cannot account for mind as it cuts this latter off from its socio-communicative dimension (i.e. relations with other minds); *the very features that make a mind a mind and not a brain* to state things in a 'folkish' manner. A few months later in that same year, Brian Scassellati from Yale University (at MIT AI Lab. at the time) used the following citation from Turing's famous article presumably in order to sum up his Doctoral Dissertation (first citation, placed top centre-page, Chapter 1).

Instead of trying to produce a program to simulate the adult mind, why not rather try to produce one which simulates the child's? (Scassellati 2001)

I do not have the impression that exponential progress in the area of "humanoid robotics" has overcome the philosophical hurdle to capture the *dialogical* essence of mind that Turing himself was aware of fifty-five years ago. With his "embodied theory of mind" Scassellati may have been referring to – or taking inspiration from – works such as Jordan Zlatev's 1997 well-written work on *Situated Embodiment.* And could it be that Scassellati would agree with Milner's limits (*cf.* supra) on explaining human cognition with the concept of learning?

Whatever the relation, academics working in Robotics and related fields like Human-Machine Interaction and Artificial Intelligence often seem to undergo an out-of-proportion positivistic enthusiasm for their 'babies'. Why is this? Don't any of them have the liberty to *really* express their doubts? There surely must be some conceptual hesitation in their mind when the action implied by their work constitutes replacing human beings. Fortunately, when they do replace a human being with a machine, it is quite often in the context of repetitive task handling that human beings no longer like to do. But there are a few academics that work on challenges that remain purely technological in nature (i.e. not that useful since man does not want to give up the action concerned – examples involving speaking come to mind). Their technological audacity does not stem from usability reports or interviews with users. Simply defying the laws of nature is what they seek to do.

Scassellati gets his expectations about *machine intentionality* the wrong way around when he writes about the "Implications to Social Robotics" of his work:

> Rather than requiring users to learn some esoteric and exact programming language or interface, more and more systems are beginning to use the natural social interfaces that people use with each other. People continuously use this extremely rich and complex communication mechanism with seemingly little effort. The desire to have technologies that are responsive to these same social cues will continue to drive the development of systems [...] Theory of mind skills will be central to any technology that interacts with people. People attribute beliefs, goals, and desires to other agents so readily and naturally that it is extremely difficult for them to interact without using these skills. They will expect technology to do the same. (Scassellati 2001, 159)

In fact, interlocutors in human-resembling communication like to be reassured that their interlocutor is human. If one wishes to escape from the Electrical Engineering and Computer Science point of view, one has to read for example the works of D. Norman, a cognitivist who addressed the DARPA/NSF Conference on Human-Robot Interaction in... yes, the year 2001. He then gave an analogy to persuade any human being to understand why *machine speech should not be flawless* in the human sense[2]. And he is not the only one that argues this point on the same basis (*cf. infra*).

---

[2] After exposing a version of the Asimovian laws of robotics, he states the following: "while speech input is still imperfect, the robot must make this clear [...]." He them gives the maxims; the first of which is: "Don't have flawless, complex speech output at a level far more sophisticated than can be understood. If

Brain-child projects are fine, but will they ever lead to a "mind-child"? Perhaps this term was never coined, in English anyway, because this notion is out of reach (whereas that of brain-child is). We should look at the further specialised field of Robotics and Computation. At least one influential author has caught my eye.

"Artificial Problems"

Some authors like to delve into "thought experiments" using 'funny' examples to study the possibilities of resolving some problem in the Artificial Sciences. Let us try to understand, in simple terms, what J. Zlatev meant in his (yes again!) 2001 article in *Minds and Machines*. His goal was to use one of these "thought experiments" in order to upgrade the position of the Artificial – robots – on the social status scale, or perhaps quite possibly, to argue in favour of taking robotic technology even further ahead. Or was it only to test the plausibility of lifting them up to our level?

In any event, he devises a fictive situation for this purpose. A two-year old child is sitting on the floor and interacting with his father through eye contact as they pass things likes balls and blocks back and forth. The child gestures towards an object that is out of reach and says "*train*". Dad says "Oh, you want the *train-engine*". In receiving it, the child repeats "*train-engine*", thereby indicating that the adult's slight correction concerning the proper term of reference has not passed unnoticed; etc. etc. (Zlatev 2001, 155). Zlatev then tells us that, when it comes to playing simple language games like this, you can remove the two-year old and put a robot in the same spot on the floor to occupy Dad; he says that today we can build a robot that would have the same physical and intellectual capacities as this person's son or daughter. I agree with him so far.

My endeavour is to focus on the communication part of his proposal as I believe this is where robotics would basically stand to gain the most from my critique.

As communication is a social activity that does not have anything really to do with physical entities or genes themselves, I am sure the

---

the robot wants people to realise it has imperfect understanding of language, it should exhibit these imperfections in the way it speaks. (If a foreign speaking person could speak fluent English but only understand pidgin speech, the more it spoke flawlessly, the less other people would understand the need to speak in pidgin)". *Cf.* Norman, D. (2001), "How Might Humans Interact with Robots? Human-Robot Interaction and the Laws of Robotology", keynote address, *The DARPA/NSF Conference on Human-Robot Interaction*, San Luis Obispo CA, September.

author pointed out here will have no objection: he does in fact carry his point of view well outside of the materialistic topics traditionally spoken about in robotics.

One does not really have to read beyond the Introduction of Zlatev's rather lengthy article (though it does read quite nicely) to find out whether his "Epigenetic Robotics" will not be able to defy the tormenting philosophical questions that Strong AI has been battling with since the days of R. Schank in the 70s, Herbert Simon and A. Newell at Rand Corporation and CMU in the early and mid 50s and still yet others; namely questions such as "is it possible for Man to build a machine to think?". As Zlatev (2001, 157), I do not have the philosophical wherewithal, but esteem myself to be able to bring a certain number of issues to his attention, though only really in point form. What I have to say could be very important for specialists in Robotics and "natural computation".

I understand epigenetics to be a field of study that involves mainly the "physicalist options" of the Cognitive Sciences; the work of Zlatev and Dennett are encouraging as they do endeavour to look into the other options possible under this banner, even if the latter author mentioned here has confused the notions of *mind* and *brain* in the past (Denett 1996).

My Approach

My vision of the way things are for the sciences of the Artificial in general, and thus natural computation and Robotics in particular, will quite simply be based on the two questions brought forth by the author:

1. If a robot is able to participate in simple language games as adequately as a child, should we ascribe true meaning and intelligence to it?
2. How would we go about developing a robot which could possibly live up to a positive answer to the first question?

My approach, slightly modified now, is still straightforward:

a. admit the first purely on the basis of verisimilitude (thus behaviour only) but refute it in the basis of veracity;

b. forget the last. In order to not leave specialists in robotics following the example targeted here in the dark, I will

c. deploy a prospective epistemology which will introduce discussion leading to the reinforcement of another well-known sub-domain of AI/HCI/Robotics thought (*weak machine intelligence, cf.* the last section).

The aim is to enhance the position of the social roles that are so important to successful communication. The reader must remember that my hypothesis is that we are able to reproduce human beings to the standards set forth by veracity.

The 'How-to's of the Norsemen

If I understand correctly, what the author means by "reverse engineering" is that in recreating the behaviour of communicative intelligence, while working with the smaller units of behaviour to form the larger ones of the robot language acquisition process, the robot builder must situate his action within a long set of implications enunciated in the exact opposite order:

> Linguistic meaning presupposes shared conventions, as a form of mutual knowledge. Conventions presuppose reflexive consciousness, allowing them to be learned and followed. Self-consciousness presupposes the perception of oneself as an intentional agent. Perception of oneself as an intentional being presupposes the perception of others similarly. Hence, other-intentionality, self-intentionality, self-consciousness and language form a possibly necessary developmental progression and an artificial system aiming at real – as opposed to simulated – language use … (Zlatev, 2001, 189).

This does appear to give a more pragmatic aspect to the 'usual implementation technique' in the artificial technologies fields, but is there not something very paradoxical here?

If these are presuppositions proper, they would indicate rather that one should start by building a robot by taking the larger chunks on and then the smaller ones. In his initial *explanation* of the thought experiment involving a child playing with toys and talking with Dad, Zlatev (2001, 155-156) starts off with intentions, goes through meaning and understanding to get to the grammar part. In fact this type of discourse is typical of positivistic science that has, so to speak, bitten off more than it can chew and then wonders what to do.

At this point one may ask if robotics really does have a set methodology and direction to follow… or is it just heuristically shooting in the dark? I could even say that the (almost not) implicit goal being chased after here, recreating man in behaviour as well as social role, is so difficult that, however big the steps robotic technology is taking towards this goal (excuse the pun!), we do have a very long way to go. As I see it, a more plausible way of seeing things would be to take the larger components for the smaller ones: i.e. putting simple intentionality features of members of the human race end-to-end to build very complex grammatical constructions. *Why should the linguistic utterances of language users be considered any less complex than human intentionality?* It would seem obvious that – after saying over the last 25 years that utterances lacking their intention-driven component cannot have meaning – we could and should be able to imagine positive responses to this epistemological question. Zlatev seems to be going in the right direction but shows here the sentiment that the Robotics community will have trouble following his initiative; hence the need for an adjective in front of "robotics" on the banner (i.e. "epigenetic").

Dialogical Communication

It is clear that Zlatev's approach is based on a rather dated account of interpersonal communication. Although Grice inaugurated the discursive study of ordinary language use with his studies on "implicatures" – a welcomed advance from the area between Philosophy and 'plain linguistics'–, his (and Gazdar's) results are not sufficient for what we are expecting of robotic intelligence today. Whatever we may expect, it is entirely clear that Zlatev's model of intersubjectivity is not able to escape that of Grice's presentation pattern of intentional layering: A knows X, B knows that A knows X, A knows that B knows that A knows X, etc. (Zlatev, 2001, 182). Communication theory has come a long way since then (*cf.* Vernant, Vanderveken, Jacques, Shotter and my own works in the 90's). It has come to fully accept *asymmetry* as the basic nature of the communicative link. The layering of intentions performed by Grice would suggest that a symmetric alibi was still necessary.

The progress that has been made in communication theory could quite simply be stated as follows, though I run the risk of being accused of over-simplification:

A cannot do with B what B does with A, whatever the communicative activity is (e.g. discussing explicitly or implicitly about knowing X).

So if the author means to speak about a Robot that participates in simple language games, how can his analysis of the situation be water tight if the pragmatic nature of the relation in question here is not solid? *Who is speaking to whom?* is a simple question that resumes what I mean by the pragmatic nature of the relation and this is important as the father in the thought experiment here, as Zlatev pointed out, does not know with whom or what he is interacting. Is that really his daughter on the floor in front of him? Is it not his human daughter?

It would appear that the field of robotics is too materialist to succeed in tackling the myth of humanity. Its endeavours only represent reproducing/replacing the mere *manifestation* of communicative intelligence; the dialogical profundity of human cognition and communication skills are hard to replace, especially over and above the toddler level. This is why the AI project has been reduced. What level of dialogism can robotics-embedded AI produce? At the outset, Turing, Weizenbaum and other post-Turing discussion was about adult dialogical capacities. Zlatev tackles two-year olds (behaviour only), and Scassellati drastically gears the argumentation down once again in his work:

> The systems presented here will not begin to approach some of the complex social skills that children master even in the first year of life. (Scassellati 2001, 19)

The robotics community does seem to have some sort of awareness, comparable to that of P.M. Milner's (*cf. supra*), of the restricted benefits learning has on the final representation of a human task like communicating with others. This said, I must take this one *big* step further. In fact, there exists a logical impossibility for a robot to participate in dialogical activities because of the *primum relationis* in human communication as defined by F. Jacques as early as in the beginning of the 80s. This means that for any propositional content flow to obtain success, it is dependent on the relationship between interlocutors *that must exist prior to it*. AI *tries* things the other way around as it so far remains unable to consider the pre-imminence of the relation.

Social Status

I would have to add a few italicised characters to Zlatev's first question: If a robot is able to participate in simple language games as adequately as a child *at least in appearance*, should we concede true meaning and intelligence to it?

P. Bourdieu would say that mechanical 'objects' (like *Robot Sapiens*) are only simple artefacts, whatever that 'species' may be capable of: it is so-to-speak *Made in the Republic of Human Society* and is thus subjected to the rules therein, rules that go beyond the boundaries of mechanics, genes, synthetic flesh and other physical paraphernalia to anticipate on the intentionality of other members (Bourdieu 1982).

But what Zlatev does well in his article is point out that it is important for machines, if they are to have success in performing operations in a human way, to learn over a period of time, to have a history. They need to have the opportunity to acquire the skills to make their own knowledge evolve. I think it is safe to say now that the programmed-in *only* method is 'out' since people working in robotic have started to take into account, as Zlatev, the more philosophical-like discourse being produced on their subject (Dennett, Dreyfus, Turkle, etc.). But the inherent phenomenon of age reduction in the object of simulation over the years does show the weakness of Man before the monumental task of explaining mind. One cannot simulate it without its explanation. It also shows the meagre contribution of the approach based on learning. Who knows, might it have a negative affect? If the ultimate goal really is to recreate humans, then the social role of learning does need to be integrated into the enterprise, but perhaps in stating its limits with respect to transforming the materials involved. Again, one may take P.M. Milner's analogies with animals for what they are worth:

> It is foolish to hope to understand the behaviour of higher animals such as primates, while remaining ignorant of the instinctive mechanisms that are only partially, and often with great difficulty, modified by learned adaptations. Consequently, before tackling the more difficult problems that arise when innate behaviour is augmented by learning it is helpful to try to understand the behaviour of animals that can learn very little. (Milner 1999, 7)

Robots and Punishment?

One of the main points I aim to point out is that society is far from being in a position to accept the *advanced* products that come of robotics, even if neuro-robotics is making good progress now. These products *are* for our utilitarian society but, in order to be fully accepted by the Self – as is the Other in a dialogical setting –, without the proper *identity features* they will remain at the fringe of human communities. Zlatev finds it necessary to play with our emotions to get his point across and so speaks of the remembrance of persons dear to oneself while they are in a deceased state (*cf.* the second thought experiment at p. 160-161); it is however a rather good idea to use strong emotions – they enhance argumentation. Think about a young boy, say in the 5-10 year-old range, who comes into the living room to alert his parents of some happening and, in the middle of their discussion, our eavesdropping reveals the following utterance:

"The robot is bothering Betty".

Listening along, we conclude that Betty is the boy's two-year old sister. Would the parents react in the *usual manner*? That is to say in the same way as when Betty's human sibling standing before them bothers the two-year old? Would the parent regularly "commissioned" to handle such a scenario go into the recreation room with the intent to, say, scold the robot?

The robot will not possess the necessary proper *identity features* in our society for some time yet to receive the treatment that might habitually correspond to bothering, teasing, pushing, hitting and so forth. For example, I doubt that, even in ten years time, scolding robots would come into practice – oh, and taking one over one's knee, even less so.

By "proper identity features" for a robot to function in a normal way at a societal level, I quite simply refer to *social status, family-induced selfhood* and *moral existence*, features *perceived as so by humans*. Furthermore, purely logical reasons for artefacts like robots not being equipped for total integration into human society do exist in the literature in Human Sciences, such as those stated early pertaining to the *pragmatic* aspects of communication: I demonstrate this in an in-depth manner elsewhere (Schmidt 2006).

Robotic "Intelligence" T*weak*ed down for Parents and other Adults

I once formulated a question for the humanoid robotics community "Can simulating Man's physical abilities meet up to the expectations we have of Robot Technology?" (*cf.* forthcoming article in *Minds and Machines*). The fact that T. Watanabe *et al* "abandoned" their InterRobot (iRT) technology for a "lesser embodied" form of communicative interaction with humans – iRT's on-screen version called InterActor – is indicative of the difficulties human speakers experience in interacting with very similar-looking creatures (Watanabe 2003, 430-435). Of course, writers as influential as H.L. Dreyfus (1972) have strongly suggested that the lack of corporal extension was the hindrance computer programmers met up against in the project of simulating human intelligence, but it has been proven both experimentally and argumentatively that fully simulating natural language (Luzzati 1989) and full simulation of human features (Schmidt 2001, 2004, 2005) goes against all sensible *logos* to improve interaction despite the more advanced level, i.e. adult interaction. This goes along quite well with D. Norman's well-fitting 'law-like' advice for designers about the flawlessness of machines mentioned earlier.

The interface, whatever the form, can get in the way; this is even more so the case for surprisingly human forms (a Cog, InterRobot, Kismit, Asimo or HOAP -type creatures do get one's attention). I argue therefore for the use of *Weak AI, Reduced Robotics* and *Invisible Interfaces.* Tweaking the scientific community towards using the "weaker" approaches *is* necessary for producing useful robotics. Taking research in this area back up to the adult level is the main idea, is it not? Perhaps it should not be? In the words of K. Dautenhahn,

> [...] humanising technology does not necessarily require creation of humanoid technology, it could rather push forward to develop technology which meets the specifically human ways and strategies of (socially) living and surviving. (Dautenhahn 1997, 33-43)

Tweaking the intelligence of neuro-robotic contraptions down into an acceptable form might mean that we need to redefine what "humanising" should mean. Could it be that developing totally human learning and development skills in the machine is not what humanising technology means? If one looks at this question with a GOFAI background (Good Ol' fashioned Artificial Intelligence), I am sure that is what it does mean. And because the attribution of human-like intelligence to technical objects has implicitly been driven into to the minds of the general public over the last half a century by the sensationalism of Artificial Intelligence, human-like robots, etc. taken up in the media, "humanising technology" does have the same meaning for all now. Perhaps the most interesting questions to be asked at this point are not those of definitions, but those that point towards utilitarianism and or the usability of these neuro-robotic contraptions.

The example given in the last section about scolding a robot (try and refrain from laughing) evokes a strange feeling concerning the falsity of the situation. Why is this? Will we, in the near future, be unable to scold neuro-robotic contraptions as one would a human youngster? The situation being somewhat comical, it is of no surprise that a mother or father cannot help but wonder what good scolding a "mechanical being" would do. Now take a look at the one-being-scolded's side of things. Is the moral dimension of a human that leads to shame and fright in such a situation present? *Can* it be, if programmed in? Or learnt in? Or both? Knowing what is happening "in" the mind of the other has always been a tough problem for philosophers. The questions concerning the very utility of the present technology comes forth as one wonders what would happen if the eye of the beholder of the entity was not aware of its artificiality. This could/would cause communication difficulties as I have discussed in much of my previous research; over and above this severe concern for the user of the technological object, the neuro-robotic contraption, the meaning of handing out social roles to objects demands justification. Has no one ever thought robotic technology would bring us this far? All one has to do is imaging neuro-robotics as a scientific test-bed being merged with other human-related technologies like the genome project, transhumanism and so forth and we would gain a situation in which we humans would ascribe full-human intentionality to the artificial with great ease without even knowing it – wouldn't we then not wish to know? The scientist that will definitively create such a "wonder" would have to explain why he is going about this. Why force man-made objects to live like humans?

Perhaps I do not fully understand. Perhaps these entities are not meant to be human. If this is the case, why do they look so much like us?

***

In this article, my intention was to augment the attention humans allot to the dialogical aspect of their lives; this is meant to combat the individualism that leads to absurd uses of scientific knowledge. I sought to challenge the widely approved idea of simulating phenomena from the realm of human mental and physical life by applying a post-cognitivist principle of mind, dialogism, to one of Man's obsessions, artefactualising his very person and experience. For there is a trade-off: in the process of "bringing matter to life", humans conceptually reduce the living down to *token simulations* of life. Is society ready to accept this? Only going *explicitly and consciously* back to the source of dialogism can help avoid the possible distortions of reality. We go to this source regularly, often in an implicit manner, i.e. accepting proper noun exchange "I"/"you" with the Other – in everyday conversation, in thinking of a loved one, in reading, etc.

Might a happy new parent one night wake up and wonder whether the robotic boy s/he "adopted" – having forthrightly scolded him/it in the day time – was truly a real person? Would it matter? This depends on whether the treatment one reserves for robotic children is the same as for biological ones. Like questions, very strange indeed, will inevitably haunt us in the future if we do not give the acceptability issues of technology priority over the mere technical challenges invented in the name of Science.

HUMAN ENGINEERING: A PHILOSOPHICAL RESPONSE

KURUVILLA PANDIKATTU

Introduction

As Paul Ricoeur observed, "every understanding is self-understanding" (Pandikattu,), and so this essay is precisely a humble attempt to understand human beings and the human situation in the present technological world.

Here we take up indirectly the fundamental human questions:
Who am I?
What can I know and achieve?
What can I hope for?
Then we look at the contemporary human technology reflectively.

The specific topic I study in this essay is genetic engineering; an area where there is a great deal of public interest and concern with regards to its profound effects. Some have raised alarm while others have welcomed it as a representing "a brave new world". We attempt to consider some of the crucial challenges as well as the opportunities of this genetic engineering.

In the long history of humanity, the discovery and utilization of fire was a tremendous achievement (Pandikattu, 2004, 19-24). Fire has not only provided humans with light and heat, but has led humans to shape the natural resources into a world of human utility. It has enabled humans to shift from muscle power to mental power by changing nature. Much more than fire, genetic engineering and prowess will alter human destiny. In a way genetic prowess give us the "inner fire" that will goad life from within and guide the destiny of both human and biotic existence.

*

In this essay, after introducing human engineering, I take up some of the crucial promises and perils offered by this great technology. Before concluding, I take up some philosophical investigations, so as to appreciate human beings and their destiny more authentically. My basic premise in this article is that, although the human being is not exactly an artefact, artefacts are found only in the human world and not in the animal or divine world. Therefore transhumanism (Transhumanism, 2009) lies at the boundary between natural and artefact.

2. Human Engineering: Basic Insights

Today it is no exaggeration to claim that scientists are beginning to understand as well as reorganize life at the basic or genetiç level. The promise and perils of this revolution are unparallel in history. In order to understand its application to human life and to study the possibility of human engineering, we may hold that four convergent forces that ushered in the genetic revolution and the possibility of human engineering are:

1. The Science of Genetics
2. The Computer Revolution and DNA
3. The Tissue Culture
4. The Economic Factors

We shall deal with each of them at a preliminary level, so as to dwell on its promises and perils.

2.1. The Science of Genetics

In fact we may trace the roots of genetic science to the early Greek philosophers. Plato the great Greek philosopher accepts a kind of heredity. Aristotle also seems to believe that the 'concept' of chicken is implicit in the egg, or that acorn was 'informed' by the plan of the oak tree. But it is only with the rediscovery of the great work of Austrian monk Gregor Mendel in 1863 that the new science of genetics developed. By observing and experimenting on garden peas, he was able to study the various characteristics and how they are passed on from one generation to the next. He made the distinction between the dominant and recessive traits which helped one to understand the phenomenon of transmission of characteristics.

This led some scientists to suggest means of improving human species, or eugenics. Francis Galton, a cousin of Charles Darwin championed the movement that aimed at improving the human race by applying the laws of heredity. He and his sympathizers called on the Government to prevent the propagation of the 'unfit' humans by forbidding their marriage, separating them from the society or forcibly sterilizing them.

Some scholars and politicians saw eugenics as a salvation for humanity from poverty, crime and other social evils. The eugenic movement reached its climax in 1924 in the United States and tumbled down with the crash of the stock market which rendered many of the elite on level with the poor. The year 1933 saw the rise of Hitler in Germany and soon he enacted the Hereditary Health Law, a eugenics sterilization statute; a great eugenic campaign. Indeed the Nazis used arguments from eugenics to justify many of their atrocities.

A better understanding of transmission of heredity through DNA took place after 1940.

Oswald Avery experimentally defined the role of DNA as the genetic material. Further, the discovery of the structure of DNA by James Watson and Francis Crick in 1953 provided the stimulus for the growth of genetics at the molecular level and one saw a period of intense activity and excitement as the main features of the gene and its expression were determined. Its triumph was the announcement of the completion of the human genome project in February 2001, by the privately funded Celera Genomics and the publicly funded Human Genome Project.

2.2. Computer Revolution and DNA

After more than forty years of running on parallel tracks, the information and life science have fused together into a single technological and economic force. Computers are increasingly being used to decipher, manage and organize the vast genetic information that is a raw resource of the emerging biotech economy.

As a result, bioinformatics has emerged as a new discipline. Scientists are now able to catalogue rich genetic information in the new genre of biological data banks. The symbiotic marriage between computers and genes has brought about new store houses of genetic capital for the use of the biotech industry. The gigantic emergence of this phenomenon could be studied under the following subtitles:

Manipulation of the DNA: Historically, it was in 1967 that the enzyme DNA ligase was isolated. It was found that this enzyme could join two strands of DNA together, a prerequisite for the construction of recombinant molecules, and is sometime regarded as a molecular glue. The isolation of the first restriction enzyme took place in 1970. Restriction enzymes are nothing but molecular scissors, that cut DNA at a precisely defined sequence.

Stanford University generated the first recombinant DNA as early as 1972. For the first time the scientific community realized that scientists could now join DNA strands together and could link the DNA of one organism to that of a completely different organism.

In 1973 scientists made yet another leap in this field when they successfully joined DNA fragments to the plasmid pSC101, which is an extrachromosal element or plasma gene isolated from the bacterium *Escherichia coli.* These recombinant molecules could replicate when introduced into *E. coli* cells.

Marriage of Computers and Genes: Several researchers are already engaged in mapping and sequencing the entire genome of organisms from the lowliest bacteria to human beings with the goal of harnessing and exploiting genetic information for economic purposes. By the end of the twenty first century, the molecular biologists hope to have the genetic 'blueprints' of tens of thousands of organisms that populate the earth. This biological information is so great that it can only be managed and stored electronically in thousands of databases in computers.

The successful cataloguing of the human genome, mentioned already, has demonstrated the need for the power of the nexus between life sciences and computer sciences. Mapping and sequencing genomes is only a beginning.

Understanding and chronicling the webs of relationships between genes, tissues, organs, organisms and external environment, the perturbations that trigger genetic mutations and phenotypical responses, is heavily depended on the computational skills of the information scientists. As a result bioinformatics has come of age. Computer titans like Bill Gates and Wall Street pundits are pumping a huge amount of funds into bioinformatics.

Virtual Biological Environments: Still significantly, today computers are being used to generate virtual biological environments to study biological organisms, networks and ecosystems. These virtual environments allow researchers to create new hypotheses and scenarios that can be used in the laboratory to test new agricultural and pharmaceutical products and medical treatments on living organisms.

Working in virtual environments, biologists can create new synthetic molecules with few strokes, often bypassing the laborious process that can often take years of effort on the lab bench. The compound, known as

QM212, was generated in the computer and its real-life counterpart is now being batch produced in several biotech laboratories.

Scientists plan to generate all sorts of new compounds that could reproduce themselves, conduct electricity, detect pollution, stops tumours, counter the effect of cocaine and block the progress of AIDS in the near future.

DNA Chip and Molecular Computer: In 1996 molecular biology took everyone by surprise with the announcement of the first DNA Chip. These chips resemble computer chips and are packed with DNA and are designed to read the genomic information in the genomes of living organisms. Some scientists use them to detect genetic deformities.

Scientists claim that in the near future the DNA chips would be able to scan an individual patient, read his or her genetic makeup and would be even able to detect genes that function abnormally. Scientists say that we shall be able to detect which genes flick on or off at any given time (TOI, 2009, 1).

The final frontier in the integration of the life sciences and information technology comes in the form of molecular computer, a thinking machine made of DNA strands rather than silicon. Scientists have already developed the first molecular computer and most of them feel that these are the computers of the future.

Unlike most computers that are sequential and can only handle one thing at a time. The DNA computers on the other hand are massive parallel computing machines that theoretically compute a hundred million billion things at once.

To sum up: the emergence of powerful computers and its coupling with biotechnology has led to the present astounding progress in bioinfomatics.

2.3. The Tissue Culture

Some scholars date the growth of tissue culture to as early as 1885. It is reported that a German embryologist Wilhelm Roux succeeded in maintaining the medullary plate of a chick in a warm saline medium for three days. In 1903, Justin Jolly, a histologist at the College de France, made a careful observation on *in Vitro* cell survival and cell divisions using salamander leucocytes (*linkinghub 2008)*. In the early experiments, fragments of plant parts such as leaf, stem anther, bud, embryo, etc., were cultured under controlled and aseptic conditions, and this technique came to be known as tissue culture.

One of the main difficulties of tissue culture was to keep the tissues free from contaminants. Thanks to the work of Alex Carrel, aseptic techniques came to be used in tissue culture. Due to the nutritional needs of cells, embryo extracts or animal blood serum came to be added to the culture. These were vulnerable to contaminants but the addition of the antibiotics, penicillin and streptomycin to the culture from the 1940's onwards alleviated this problem.

Another significant development was the use of trypsin (a proteolytic enzyme) by F.S. Rous and P. Jones in 1916 to free cells from callus tissue. Today the tissue culture is much advanced with standardized culture media and sophisticated incubation conditions. By the 1940s and 1950s, tissue culture media were developed and techniques were worked out that closely simulated the situation *in Vivo*.

Ordinarily, tissue culture is divided into two: (1) Organ culture, (2) Cell culture. In organ culture, whole embryo or small tissue fragments are cultured in such a way that they keep their tissue totipotency. Cell cultures on the other hand are obtained either by enzymatic treatment or mechanical dispersal of tissue into individual cells or by spontaneous migration of cells from an explant, and they are maintained as attached monolayers or as cell suspension. Freshly isolated cell cultures are known as primary cell cultures. Once a primary culture is sub-cultured, we get cell lines and when a complete animal is obtained from a somatic cell of an animal, it is christened as animal cloning. Thus we have the cloning of Dolly, a sheep in 1997 at the Roslin Institute in the U.K by Ian Wilmut, Keith Campbell and colleagues.

2.4. The Economic Factors

Genes appear to have become 'green gold' and one can already notice that the political as well as the economic powers are all out to gain control over the genetic resources of our planet. Hence multinational corporations are funding research and are scouting the continents in search of genes that have market value, with an eye of profit.

Commoditisation of the gene pool has lead to their patenting. Thus, letting the control of our life into the hands of the unscrupulous business masters who might commercialize it for mere profit, is a real danger.

The issue of patents and profits derived from it are serious related concerns for humanity. In 1971 an Indian microbiologist, Ananda Chakrabarty an employee at General Electronic Company applied to U.S. Patent and Trademark Office (PTO) for a patent on a genetically engineered microorganism designed to consume oil spills on the oceans. PTO rejected while the court of customs approved it. In 1980, Chakraborty won by a slim margin of five to four in the Supreme Court.

Genetic engineering is about turning genes, the heritage of millions of years of evolution and their conversion, into intellectual property of some corporation that in the name of protection from bio-piracy end up with monopoly over the genes. This commoditisation and privatization of life has profound ethical and philosophical implications.

3. The Promises and Perils of Genetic Engineering

Genetic Engineering has the power to bring about a new genesis on our planet. The possible wholesale transfer of genes from totally unrelated species and across all biological boundaries, plant, animal and human, and the resultant emergence of thousands of novel life forms in a brief moment of evolutionary time chiefly controlled by economic gain is not beyond the iota of abuse.

3.1. Genetic Engineering and Agribusiness

Genetic Engineering is mostly used commercially in the agricultural sector. Plants are genetically modified to have an inbuilt resistance to pests and to fix the atmospheric nitrogen like the symbiotic bacteria Acetobactor. Insects are genetically engineered to attack the crop predators.

C.S Prakash, the director of the Center for Plant Biotechnology Research at Tuskegee University in Alabama, and his team of researchers have developed a sweet potato with five times the amount of protein of a normal sweet potato. Such an improvement can bless millions of people where the tubers form the staple element of their diet.

Terminator seeds that claim to give us high yield are being designed by multi-national companies like Monsanto. These seeds are designed to be sterile unless activated by a chemical that only the company sell. This move allows them to control genetic pollution as there might be possibilities of cross-pollination with weedy relatives creating super-weeds.

Some scientists say that genetically modified organisms provide us the best opportunity to feed approximately 800 million people who now are victims of malnutrition. But the use of genetically modified food for human consumption does raise questions of human health and well being. Some of the significant areas of concern in the agribusiness are listed below:

Genetic Pollution: Every genetically modified organism that is released in the environment poses a potential threat to the ecosystem. The genetically induced pollution is very different and inherently more unpredictable than the petro-chemicals in the way they interact with the environment.

Genetically engineered organisms are self-replicating. They grow and they migrate. As a result it is unlike the petro-chemical products, it is difficult to confine them in some geographical place. Once released in the environment it is next to impossible to recall them back into the laboratory, especially those organisms that are microscopic in nature.

Much of the research in agricultural biotechnology is centred on the creation of herbicide-tolerant, pest-resistant and virus resistant transgenic plants.

Herbicide-tolerant plants: Overuse of herbicides due to weeds developing resistance, may result in causing great harm to us, soil, water, and beneficial insects.

Pest-resistant plants: Overuse may lead to the grow of 'super bugs'

Virus resistant: Overuse may lead to the emergence of new viruses that were never known before.

Finally, Gene flow involves the transfer of the transgenic genes from transgenic crops to their weedy relatives by way of cross-pollination. This also brings up the dangers of gene flow of herbicide-tolerant, pest-resistant, and virus resistant transgenic super weeds.

The ambitious plans to engineer transgenic plants to serve pharmaceutical factories for the production of chemicals and drugs, vaccines and industrial enzymes subject many seed-eating birds, insects etc., to serious risks of untold consequences.

3.2. Stem Cell Research and Possible Human Cloning

Stem cells promise huge therapeutic benefits to humans. Stem cells have the ability of developing into needed body parts such as tissues and organs. Stem cells are chiefly harnessed from embryos and this results in abortion and hence remains questionable. But stem cells obtained from adult skin, blood, and bone marrow, and fat and blood harvested from the umbilical cord is highly recommended.

The goal of any stem cell therapy is to repair any damaged tissue that cannot heal itself. This might be accomplished by transplanting stem cells into the damaged area and directing them to grow into a new, healthy tissue. It may also be possible to coax stem cells already in the body to work overtime and produce new tissue.

To date, researchers have found more success with the first method, stem cell transplants

An Indian centre has reported a breakthrough in March 2009. It is a breakthrough that may have the country's medical and scientific community sitting up and taking notice. The Centre for Stem Cell Research at Christian Medical College, Vellore, Tamil Nadu, has succeeded in reprogramming cells drawn from adult mice and making them function like stem cells found in the human embryo (TOI, 2008).

The successful cloning of Dolly in 1997 and the possibility of human cloning has provoked a worldwide debate. Subsequent reports that the Hawaiian scientist producing three generations of mice and the Japanese scientists producing eight identical calves from a single adult cow has brought us closer to the possibility of cloning humans.

Some predicted: "Today sheep and tomorrow the shepherd." Soon a French scientist, Brigitte Boisselers, the president of cloning society Clonaid, claimed that her team had cloned a baby girl whom she christened 'Eve'. There was a hoax claim from a Korean scientist that he had cloned humans. Many sober scientists rubbished their claims. Yet we all know that humanity is at the threshold of human cloning.[1]

Cloning basically involves an implantation of the nucleus into a de-nucleated egg which then is chemically treated so that the combination behaves like a fertilized egg (zygote). This fertilized egg develops into an embryo with the entire genetic code of the implanted nucleus. The issue of cloning is of urgent importance as it represents a radical break from the past and can affect our future in a deeply significant way. Although it has

[1] Washington, Mar 10 (PTI) President Barack Obama has said he will not allow human cloning in US even as he lifted the eight-year-old ban on funding of stem cell research. "We will ensure that our government never opens the door to the use of cloning for human reproduction" Obama said in his remarks, before signing the executive order to lift the ban on funding of stem cell research. "It is dangerous, profoundly wrong, and has no place in our society, or any society" Obama said as he promised to enforce strict regulations with regard to stem cell research in the country. Promising that his Administration would never undertake this research lightly, Obama said: "We will support it only when it is both scientifically worthy and responsibly conducted. We will develop strict guidelines, which we will rigorously enforce, because we cannot ever tolerate misuse or abuse." (Obama, 2009)

many therapeutic boons, the spectre of it being misused is equally alarming (Pence, 1998, 168).

The possibility of the germline manipulations give us the power to manipulate the fertilized egg, the first cell in the embryo, so that genetic changes could be copied in every cell of the future adult including his or her reproductive cells. (Stock & Campbell, 2000, 79 & 85)

This lead to the possibility of transforming the basic chromosome in the human body and making ourselves "GeneRich" individuals. Limited only to a few elites, this has the possibility of paving the way to another better or improved human species, as already predicted by Lee Silver's Remaking Eden is such an attempt part fictional, part scientific (1977).[2]

Published in 16 languages, *Remaking Eden* introduced the concept of "reprogenetics": the combined use of reproductive and genetic technologies not for the treatment of an existing medical affliction, but to enable prospective parents to choose which genes their children receive. Many potential reprogenetic applications could be viewed in a positive light when considered in terms of their impact on individual health and well-being.

However, *Remaking Eden* also spins out a speculative scenario of a dystopic future - with the divergence of genetically enhanced *GeneRich* and unenhanced *naturals* into separate species - as a direct but unintended consequence of *human nature* and the principles of liberal democracy.

3.3. Our Collective Destiny

Confronted with the radical future promised by genetic engineering, we can visualize our collective future either in terms of progressive transhumanism (Transhumanism, 2009) or total self-destruction.

Transhumanism: The possible engineering of human beings is not limited to remedies, or rectifying the defects in the human body, but to transform human nature radically, as indicated by movements like World Transhumanist Association and The Humanity+.

Such organizations advocate the "ethical use of technology to expand human capacities." To quote from their website: "We support the development of and access to new technologies that enable everyone to enjoy better minds, better bodies and better lives. In other words, we want

[2] Published already in 1977, the book has the subtitle How Genetic Engineering and Cloning will Transform the American Family Harper Perennial. Another version of the book is: L. Lee Silver. *Remaking Eden: How Cloning and Beyond Will Change the Human Family* (New York: Avon Books), 1977.

people to be better than well." This leads to the possibility of a new species: the transhumans (Hayes, 2009).

Transhumanism is a way of thinking about the future that is based on the premise that the human species in its current form does not represent the end of our development but rather a comparatively early phase. It is formally defined as: "The intellectual and cultural movement that affirms the possibility and desirability of fundamentally improving the human condition through applied reason, especially by developing and making widely available technologies to eliminate aging and to greatly enhance human intellectual, physical, and psychological capacities." (Transhumanism.org)

Posthumans is the next expected phase of human evolution and is used to talk about possible future beings whose basic capacities so radically exceed those of present humans as to be no longer termed unambiguously human. The standard word for such beings is "posthuman" as opposed to "transhumans." (Fukuyama, 199, 28 and 33). Many of the proponents of this school believe that humanity is on the verge of evolving into transhumans!

Total Destruction of Human Life: The scenario presented above gives either a gloomy or hopeful perspective. Either we can rise above our human limitations and go to another level of existence or we lose our valuable humanness and become "super-humans," depending on our perspective. Whatever the "normal human life" is as we know it, will most probably fade away if the transhumans emerge.

On the other hand, there is the other serious danger of us truly messing up with our own existence, using the very means meant to elevate us. We can use the same technological potential at our disposal to eliminate human life altogether.

Genetic warfare, making use of micro-organisms like bacteria, viruses and fungi for military purposes, provides one means for total destruction of human life. The knowledge accumulated through genetic engineering may be used to develop a wide range of pathogens to attack plant, animal, and human populations. Biological weapons can be viral, bacterial, fungal and protozoan. These biological agents can mutate, reproduce, multiply and spread over large geographic terrain by wind, water, insect, animal, and human transmission.

The recombinant DNA technology allows the possibility of creating nearly an infinite variety of designer pathogens that were never seen before. It is possible to insert lethal genes into otherwise harmless microorganisms. This research is also laden with dangers of accidental intrusion of designer gene weapons from the laboratories into our environment causing great damage to us and our ecosystem.[3]

Therefore it is up to us to act on the human destiny awaiting us: a radically transformed super human species or the possibility of the extinction of ourselves.

4. Philosophical Analysis

Confronted with the crucial choices presented above, we need to reflect on our own existence, our responsibility to each other, including the larger environment and to our future generation. This takes us to the final section when we reflect on the uniqueness of human existence, on our moral life and on the larger issues like life and its evolution. Obviously the issues raised have to be very limited.

4.1. Anthropological Issues

The crucial question that confronts us as humans collectively is: Who are we? Going beyond our individual identities, now the crucial issue is that of our collective identity. It is well-established that humans are social beings. Much more than our social nature, the issue at stake is: What is human nature? Since by our very nature, we are called to go beyond ourselves and transcend all barriers, our very nature is in a state of permanent enhancement. So the question to be asked is: How emergent and pliable is human nature?

Closely related to this issue is the question: What is human life and who "owns" it? Who can be responsible for the collective responsibility of our own destiny? Can we entrust it to our politicians, scientists or corporations? The wider issue is: who speaks for the earth? Or who really can be our representative, when the larger issue is human life itself?

At the most crucial point in human history, when we are about to bid goodbye to ourselves, who can authentically and responsibly represent us and take life-shaking decisions for the rest of us? Obviously it is not up to a few individuals, but a group of concerned, compassionate and knowledgeable citizens! But who are they? Whom can we entrust the whole of human destiny to?

---

[3] For the time being we do not consider other possible ways of humans destroying or eliminating themselves, like nuclear weapons, climatic degradation, global warming, violence or hunger leading to social catastrophe.

4.2. Moral Concerns

Closely connected to the larger anthropological issues are the moral concerns. There are very many legitimate questions on the use of various techniques like stem-cell research, public funding, profit motive etc.

Leaving aside these issues I want to take up only two moral issues: Who can legitimately decide to evolve into transhumans or posthumans? Is it limited only to the elite who can financially and heuristically afford it? Who chooses these elite and how will they look upon the natural humans who are not "fortunate enough" to climb the ladder of species transformation?

We must be aware of the danger that the profit-motive of few individuals and multinational corporations take the whole humanity for a good ride! As such we cannot entrust the most crucial decision of humanity to a few privilege individuals. A still worse scenario would be if the select elite, who have attained a higher degree of evolution, look down on natural humans, just as we today look down on chimpanzees.

In fact, we can speak about the ethics applicable to the super-humans. But more importantly, we need to ask about the price we are ready to pay for ourselves. Do humans beings have an intrinsic price in themselves, which is independent of the external price one is ready to pay? As Kant suggests, if we are truly priceless, can profit alone determine the growth of humanity as a species? (Kant 2004)

4.3. Biotic Dimension

Much more than the few anthropological issues raised, the larger biotic issue is the value, dignity and even ownership of life itself. Issues like who speaks for life, or who stands for the biological life, are crucial in finding a way out of the maze we have created for ourselves.

The decisions we make in human engineering shapes indirectly, but radically, life in its entirety.

So, concerns like, "Who is responsible for life?" and "Who can guarantee the continued survival of life?" must be squarely addressed. The larger question of whether we are ready to be the spokesperson for the whole living community needs to be tackled. Unfortunately I get the impression that humans have not learnt to deal with themselves and so are very poorly qualified to be the spokespersons for the larger communities of living beings.

Unfortunately our spiritual, moral and even aesthetic faculties have not kept pace with our technological progress. So we are in no position to control the very instruments we have produced. They are, therefore, threatening to control and guide us, including the totality of the life-systems. It is a serious situation, calling for urgent steps in terms of moral, spiritual and holistic response.

4.4. Spiritual Evolution

Granted that the biological (genetic) evolution so far has been able to create stupendous beings like us, should we not raise the most significant question? In spite of all our physical and technological progress, it is self-evident truth that the uniqueness of human beings is their "spiritual" qualities. The Indian categories of *sat-cit-ananda* express this development.

According to these categories, *sat* (existence or materiality) slowly evolved into *cit* (consciousness). The intelligence is a "higher" form of existence and leads further to *ananda* (bliss). In every being, these three aspects are present in various degrees. It is in the human person that they reach their maximum intensity. At the same time, there is scope for further growth in all the three levels, particularly in the last two levels.

Unfortunately human genetic engineering is at a loss to explain the evolution at the *cit* and *ananada* levels. This calls for moral and spiritual developments of human beings, which unfortunately is inadequate in the present scenario of the world community. So the crucial issue is: can people, whose moral and spiritual progress has been found wanting, be in a position to control these dimensions?

Alternatively, we also need to ask: Will the emerging transhuman species be in any way more fulfilled at the moral or spiritual levels? In other words, will they be "better" or more contended compared to normal human beings? Will they be more knowledgeable, committed and compassionate? If not, we cannot speak of progressive evolution for the transhumans.

5. Conclusion

I hold that the benefits of genetic engineering are countless and the ethical implications call us to respond carefully and critically so that the choices that we make might protect the sacredness of every form of life. We need to proceed, but proceed with care, caution, and compassion. We just cannot risk the lives of millions, nay billions, of human beings!

So on the one hand, without in any way becoming a prophet of doom, I do not in any way plead to give up genetic engineering, including possible human engineering. On the other hand, it is imperative that the world community be involved collectively in the decision we take regarding our destiny. It has to be an informed, deliberate, cautions and collective decision; one based on the sacredness, dignity and uniqueness of life. Thus we can truly look forward to a future that is truly more humane and transhuman.

CHAPTER TWO

TECHNOLOGY AND ENGINEERING UNDER PHILOSOPHERS ANALYSIS

PHILOSOPHY AND TECHNIQUE

ALEXANDRU BOBOC

1. A debate about "technique" today in the age of "worldwide technology" cannot be separated from the study of human kind or of the "human condition" in the outline of modern philosophy of culture. For technology is the field of culture and also of the expression of the productive attitude in all professions ("rationalization"): in political and social areas, in science, art, even in human emancipation in social and cultural behaviour.

In fact, the very term "technique" originating in the Greek <u>*techné*</u> (Latin: *ars*, ingenuity, skill, ability) means the way man understands himself and proposes, bringing products to fore in order to satisfy material and spiritual needs. Production (action to make something happen) associates *techné* with *poiesis* (production in general), thus associating it with the *praxis* (action in the moral sense) and with *theoria* (examination by thinking) - the ways through which man, endowed with a thinking ability, is positioned in an actively creative manner in the world.

In this sense, it becomes "classic". Technique therefore means the way and the manner to set up something, to accomplish, to achieve something, in general: as human activity is oriented (thus, of "producing-toward"), meaning is proceeding to exceed the empirically-instrumental level, what was said in the modern times to be "savoir-faire".

In a modern characterization *technique is*:

> 1. As opposed to nature: the configuration work (modelling), usufruct of materials and natural forces in the service of human needs, the implementation of goals and ideas in general and scope of activities and its products (artefacts); 2. As opposed to such activity (creative one), the formal proceed modes, the amount of means and rules to deliver such creations; 3. As opposed to art and art craft: exploring nature with machines in order to expedite work and meet the needs of people based on the knowledge of mathematical-physical science of nature, the scope of social mechanized production. (Hoffmeister, 1955, 603).

In fact, the handmade training, situations and configuration conditions, knowledge of nature, science, and understanding are the premises of *technique*, which means that as it proceeds, the *technique* occurs beyond the mere empirical-craft activity, and even beyond the power of cars in a scientific-rational organization, in a configuration that generated, in a theoretical way, the problem "technique and culture" (which was revealed, it is true, only in the twentieth century).

In these conditions, modern technology is designed to satisfy, by the expansion of industry and economy, not just the needs that arise in different areas, but also to create new needs. At the same time, the growing process of becoming technical also involves the danger of losing the authentic meaning of *technique*, affecting lifestyle, relationships between work and social conditions of completion of labour, and, not least, the human-ranking of the individual in the whole process of technical operation.

In short, the "danger" comes from the fact that the *technique* can be applied as a neutral tool in the fundamental relationship of man with nature and with his own living environment. Against this background, considerations of thinkers intervened (from the era of romanticism, but also from later on) about this "cruel devil", which is not in the service of man.

As noted:

> in fact, neither heads nor hands can change anything in the destiny of the machines technique (Spengler, 1932, 74).

and

> it belongs to those tragic times that the unshackled human thinking cannot understand its own consequences. Technique has become esoteric, as the highest mathematics, that it serves ... Mechanization of the world has entered the stage of dangerous tensions (Spengler, 1932, 78).

2. The modern world thus ascertains (with surprise and concern) that technique is more and more lacking coverage in the rational area, that it is still something hidden, secretive and, as such, difficult to explain and understand. This all the more so as "modern technology" is emancipated even in engaging in the design and goals, focusing its education strength by establishing a specific process of combining (and cooperating) with the modern science of nature.

> Engineering [Technique] and natural science are in a reciprocal relation: in the structure itself, the science of nature already seems to be technical ... conversely, the technique seems scientific; the theories themselves enter its appliances and cars, and the technical "applications" act correction requirements and concluding theories. (Ienning, 1980, 161).

Against this background there is already talk on "technical thinking": it "wants achievements. The luxury of the car is the result of the constraints of thinking";

> we live in a world marked by technique, in a techno-culture … The growing technological power of humans created through its immensity a new type of ethical situation. (Lenk, 1982, 7).

Human kind has become dependent on its own techno-culture and as a consequence "it must learn how to proceed in a wisely manner about the technique" (*Ibidem,* p. 9).

Good advice, certainly, but not so easy to follow because it all comes down "to learning" about how to behave in a world where life is made possible only with the means of modern technology.

In moderate terms (soothing, maybe) it was stated:

> Whether this "primacy of the art" is challenged or praised, appreciated, or damned, its reality is undeniable. The whole power of the shaping forces our contemporary culture is increasingly confined to this point". Even the most powerful forces opposed to technology - even those spiritual potentialities, that by content and purpose are located furthest from it, - seem to fulfil any services only by the fact that are united to it, and that are subjected to it even unnoticed in this alliance. This subordination is worth today for many as the own goal towards which modern culture is reaching, as its irresistible destiny. (Cassirer, 1930/1985, 39)

3. From the perspective of the prospect of a "philosophy of symbolic forms", "technique" is not confined to the use of machinery, installations, etc., but is

> a fundamental change in human lifestyle, in the conception on the world and on the self-understanding of it, so that the concept of "technique" changes in the place of reason; in their development, science and technology influence each other. (Krois, 1983, 69)

But both tend towards opposite ends:

for technique is a space for human life (in it, it can be created something new, making it comparable to artistic activity). (Krois, 1983, 71)

With this, technique is located in the context of "symbolic forms" (basic forms the of culture): first with art, but stating that: "technical renewals follow determined practical purposes" as opposed to art scopes technical areas "do not express human feelings" (Cassirer 1930/1985, 71).

Cassirer himself opens his analyses with the beginnings of the technique in the mythical world, understood as "the original and primary symbolic interpretation manner", but in a complex network, where a key part is played by the language function: this happens in terms of a parallelism between language and tools: here "a real philosophical problem, is hiding" both sending to a "spiritual principle", but maintaining "closeness to nature and its research". (Cassirer1930/1985, 51, 73, 74).

Here the following statement should be noted:

> technical work also agrees with the theoretical truth the basic determination: both are determined by the requirement of "correspondence" between thought and reality, an "adaequatio rei et intellectus". But in the technical creation stands out more clearly than in the theoretical knowledge that this "correlation" is not given directly, but is sought and brought to the fore gradually. The technique is subjected to nature, since it obeys its laws and sees them as being unflinching premises, but in this listening (towards the laws of nature) nature is never something over, a mere implementation (establishment), but a continuous restart, an always-forward. (Cassirer, 1930/1985, 80-81).

In essence, the question of technique

> "does not refer primarily to what is, but what can be. But this "can" (können) does not designate a simple acceptance or assumption, but it expresses an assertive affirmation and an assertive certainty - a certainty, whose last confirmation is not sought in the mere judgement (Urteilen), but the production and revealing of specific configurations. In this respect any truly original technical benefit has the features of a discovery (Ent-Deckens) as disclosure (Aufdeckens): a content actually existing in itself is thus removed from the realm of the possible and implanted in reality". (Cassirer, 1930/1985, 81)

The technics is the image/manifestation (Ebenbild) of that action that Leibniz in its metaphysics, poses in the "divine approach" that

> does not create the essentialities or the possibilities of the objects, but of all the possibilities that exist in them picks only one and that is: the most absolute. Thus, the technique offers advice on circumscribing of "that is objective", of what is objectively determined through universal laws, does not coincide with what exists, of what is made sensibly. (Cassirer, 1930/1985, 81 & 82).

4. Technical creation is never quartered in the pure "facticity", before given objects, but

> is situated under the law of a forerunner anticipation, of a vision that foresees what is outlined in the future, revealing a new future" "below the law of prior expectations, a vision that provides what is shaping the future, while putting in a new light future. (Cassirer, 1930/1985, 82).

But with this "the gravity centre of the world of technical shape" seems to move more and more from the field of theory into an artistic and creative approach. How closely intertwined the two areas are, does not actually require any evidence. "Taking a look at a general history of the spirit is enough to clarify how mobile practical theories are becoming, in the genesis of the world of technical and artistic shape". (In Cassirer, 1930/1985, 83, some references to the Renaissance are made. Thus there were found "great examples" of artistic and technical shapes in Leon Battista Alberti or Leonardo Da Vinci).

But such a kinship between technique and art should not lead to the deletion of "differences" between technical and artistic creation, which stands out by examining the way of "rendering objective" which is active in the artist and the technician:

> This is not enough to invoke the forces of nature or the forces of intellect, because we are on the point at which only the intervention of new acts of will can produce real transformation. The construction of the field of will and consciousness of what should, on which an ethical community relies upon; the technique can only be servant, not head. It cannot make aims by itself, although in their fulfilment can and should cooperate; she understands best its meaning and its *telos*, whether it recognizes it: that can never be and end in itself but only in another "area of goals", that is to integrate itself in the real and definitive teleology reason that Kant saw as ethical-teleology. (Cassirer, 1930/1985, 88 & 89).

In this regard, "dematerialization", the ethical becoming of technique is the central issue of our contemporary culture. Because the technique is under the rule of "thought, serving facts", under the ideal of work solidarity,

> It creates, even before the real community of free will, as a community of fate between all those who are active in their works. Thus the default meaning of earthly work and technical work may be designated as the thought of "freedom from servitude." (Cassirer, 1930/1985, 89).

5. The idea of action through technique as "brake of the chaotic forces in the human being" send us to the technical understanding of creation in the light of moral conscience as a common cause of it to the truth and to the needs of today's world.

Coming back to a debate in the philosophical coordinates is now all the more necessary. As in many aspects technique under many aspects linked to the industrial production overshadowed "technique" (essence of the technology), we have to reconsider the thinking on technology (beyond what was called "thinking technique").

For

> the essence of technology is by no means anything technological. Thus we shall never experience our relationship to the essence of technology so long as we merely conceive and push forward the technological, put up with it, or evade it. Everywhere we remain unfree and chained to technology, whether we passionately affirm or deny it. (Heidegger, in: Krell, 1978, 287)

The definition of technique as "instrumental",

> However, in order to be correct, this fixing by no means needs to uncover the thing in question in its essence. (Heidegger, in: Krell, 1978, 289)

> Technology is therefore no mere means. Technology is a way of revealing. If we give heed to this, then another whole realm for the essence of technology will open itself up to us. It is the realm of revealing, i.e., of truth. (Heidegger, in: Krell, 1978, 294)

For

> *technē* is the name not only for the activities and skills of the craftsman, but also for the arts of the mind and the fine arts. *Technē* belongs to bringing-forth, to poiēsis; it is something poietic. (Heidegger, in: Krell, 1978, 294)

More precisely:

> *Technē* is a mode of *alētheuein.* It reveals whatever does not bring itself forth and does not yet lie here before us, whatever can look and turn out now one way and now another … Thus what is decisive in *techne* does not lie at all in making and manipulating nor in the using of means, but rather in the aforementioned revealing ... Technology is a mode of revealing. Technology comes to presence in the realm where revealing and unconcealment take place, where *alētheia*, truth, happens. (Heidegger, in: Krell, 1978, 295)

Regarding the "threat" in the world of "modern technique" Heidegger stated:

> "What is dangerous is not technology. There is no demonry of technology, but rather there is the mystery of its essence. The essence of technology, as *a* destining of revealing, is the danger. (…) Rather, precisely the essence of technology must harbor in itself the growth of the saving power." (Heidegger, in: Krell, 1978, 295)

Thus

> "the coming to presence of technology harbors in itself what we least suspect, the possible arising of the saving power. Everything, then, depends upon this: that we ponder this arising and that, recollecting, we watch over it." (Heidegger, in: Krell, 1978, 314)

> But human reflection can ponder the fact that all saving power must be of a higher essence than what is endangered, though at the same time kindred to it. (Heidegger, in: Krell, 1978, 315)

> It is stated that to the antique not technology alone that bore the name *technē*", "but also the bringing-forth of the true into the beautiful." (p. 315) Consequently, "the decisive confrontation with it must happen in a realm that is, on the one hand, akin to the essence of technology and, on the other, fundamentally different from it. Such a realm is art. (Heidegger, in: Krell, 1978, 317)

6. Returning to what the classical tradition designed through *technē*; the art (ability, capability) to design and make something, which in modernity has been associated with the request of "practical philosophy", whose study was located mainly in the field of ethics. But this, in a Kantian new disposal, criticism in the plan of new "uses" of reason: the "technical" one, along with "theoretical reason" and "practical reason": the

nature of technique, leads beyond the simple mechanism to its concept as art";

> the technique of power of judgement is the technical reason, i.e. applied, realizing it through itself, and as well as the practice, but not as law-making, that is not pure morally- practical. (Lehmann, 1969, 355).

Kant himself wrote:

> Self-subsisting natural beauty discovers to us a *Technic of nature* (sub.n.), which represents it as a system in accordance with laws, the principle of which we do not find in the whole of our faculty of Understanding. That principle is the principle of purposiveness, in respect of the use of our Judgement in regard to phenomena; [which requires] that these must not be judged as merely belonging to nature in its purposeless mechanism, but also as belonging to something analogous to art. It, therefore, actually extends, not indeed our cognition of natural Objects, but our concept of nature; [which is now not regarded] as mere mechanism but as art. This leads to profound investigations as to the possibility of such a form. (*Critique of Judgement*, 1914, 103)

Positioning *technē* into the area of reason brings it in the circle of the forms of rationality, and the interaction with art and morality reveals a different dimension than that of the technical-science unit: human significance, which is underlined by the above mentioned developments (Cassirer, Heidegger, etc.): the technique is not a simple "means" for the acquisition of something, but a form of self-making of the human being, a form of culture.

The technical process is indeed a major way of progress, but does not fully justify the separation of the technique of the "practical philosophy", which can be found in an alienation of the technique, in the fact that technical progress is not primarily, a form of human creativity, but as something in itself (thereby hiding the so-called "threat" of modern technique).

Indeed, the essence of technique is not the same with the "technical", but understanding it requires more steps than those that came through "the task of thinking" (Heidegger), designed to actively restructure the social and human condition of technique in a civilization provided with values, into the truth, beauty and good, where "destiny" should not take the form of constraints, but the assertion of the sense of labour and creation of human personality.

The "primacy of technique" in the modern society is a fact that must be started from, but this does not exclude an emancipation through the technique itself, in a world of interaction of the forms of culture and not simply a "techno-culture", because the origin of art as technique is in humankind, not in the scale of the historical calendar, and any act of philosophising on the technique should take this into account.

"*THE*" ONTOLOGY OF TECHNOLOGY – ASSUMPTIONS AND MEANINGS

IONUȚ ISAC

To write as well as to speak about *the* ontology of technology, without quotation marks, might sound far-fetched. However, what made me choose to write it as "the" is a bi-millennial philosophical tradition inside of European philosophy, where ontology has been traditionally known as the "science of Being". Nowadays, when ontology is no more likely to be thought of in a traditional manner, a project of *the* ontology (be it the ontology of technology) as if there is only one, is questionable.

Thus, it would be more adequate to put the subject as follows: *ontology* of technology (among many possible others). This is one of the various assumptions and meanings I am trying to emphasize in my approach, being aware that the expression "an ontological model" is more adequate for such purpose. But even inside the discussions about "an" ontology of technology, there are at least two opposite points of view: one comes from the idea that ontology finds itself somehow in opposition to history; the other is the idea that ontology must not be seen as being "out of history" or *a priori* in any sense, as, for instance, Lawson writes (Lawson, 2008). I can appreciate that the ontological accounts can be better sustained having the second point of view in sight. Furthermore, a position worth noting while rethinking and assessing the ontology of technology is a *non-essentialist* or quite an *anti-essentialist* one. Coming presumably from Wittgenstein's heritage, such a point of view presumes that the reader is familiar with a variety of technologies; therefore no special theoretic treatment of the subject-matter is required. It directs the very question of defining technology by avoiding the appeal to classical definitions, according to Fellows (1995).

As any ontology of technology, aims to have within its scope (i.e. *the technical object/ system*), there are different ways to conceive their relationship. As it has been remarked, only a decade ago philosophy of technology became more analytical (in the context of an "empirical turn") about more specific and concrete technological developments. Until that time, as de Vries emphasizes (2008), this philosophical branch was rather preoccupied with much broader subjects, such as the influence of technology on society and culture. (Nevertheless, the issue remains of utmost interest). Thus, reflection upon technology has undergone a gradual "transition" from general interpretations concerning the history and philosophy of science/ technology to an analytical endeavour able to bring closer to us the parameters of the technical artefacts. I believe that this specific "pragmatic shift" proves itself to be the echo of the well-known change undergone by philosophy and metaphysics at the middle of the 20th century. Then, the classic aim of elaborating monumental classic systems gradually yielded to critical semantic analysis and pragmatic interpretations of the language.

The scope of my approach is, therefore, to discuss some of the meanings and assumptions associated with the concept of "ontology of technology", which belongs to the core of contemporary debates in the field. All of them and, of course, many others that I cannot discuss inside the present text, are important and relevant for an understanding of the problem. In the end, I shall conclude in favour of a *moderated-relativistic*, *historicist* and *non-essentialist* view upon this subject-matter, as it seems to me that it synthesizes the best current trend in ontology of technology.

*

To speak about the "ontology of technology" in a world fundamentally embedded in and almost totally depending on it seems rather easy at first sight. Like many others occurrences in everyday-life, the difficulty appears at the very moment when one tries to account for the meanings of this term. For instance, Lawson writes that:

> Not only is there a general failure to reach consensus about the meaning of the term technology, but there is often little attempt made to establish a meaning of technology at all. Indeed, many argue that technology may be seen as the archetypal black-box category of social science… It might also be argued that, until fairly recently at least, even within the philosophy of technology there has been some degree of black-boxing of technology as a result of focusing upon the social consequences of technology rather than on the nature of technology itself. However, a variety of more recent contributions have attempted to reorient the study of technology towards describing the nature of technology prior to addressing its likely effects. (Lawson, 2008)

The already existing approaches in philosophy and ontology of technology indicate that there are more meanings assigned. For instance, let us remind ourselves here that some philosophers of technology like M. Heidegger or J. Ellul have focused their ideas mainly on ontology of technology' *implications* or *entailments* (rather negative); while some sociologists and historians of technology had in mind technology understood as a *process*, i.e. *concretization* (Lawson, 2008). The first position ("monadologic") remains somehow isolated – each thinker with his own view – while the second one manages both to collect the advantages of other points of view and to avoid the shortcomings of others, as it happens with the philosophy of technology and social constructivism.

I suggest that such kind of flexible interpretation is due to the aforementioned historical shift of the philosophy of technology. Similar to what happened before in theory of knowledge and general ontology, the ontology of technology has become more focused on the concrete and historical becoming of technical items and systems. Gradually, reflexive thinking upon technology meant, to a smaller extent, the general context of human culture, as seen through a framework of the history and philosophy of science/ technology, has increasingly got the features of an analytical endeavour able to bring us closer to, and intelligibly integrate the parameters of, the technical artefacts in the general "equation" of the philosophical thoughts. Thus, case studies, historical outlooks and concrete-relativistic arguments entered and took their seats on the scene.

But this "transformation" doesn't look like an exclusive choice or an irreversible fate, since rigorous and comparative analysis does not forbid reflection on general meanings; rather they are linked or complementary to one another. Thus, it is to be expected that analysis shall provide in the future exactness to support theoretical explorations and constructions. And besides, as de Vries asserts, referring to the works of Gilbert Simondon,

> There exists some 'analytic philosophy of technology' *avant la lettre*, be it with a less developed argumentation, but still it is worthwhile to trace it back and see how it might contribute to our current interests. (de Vries, 2008)

This way, there are reasons to sustain the assumption that we cannot speak about "the" ontology of technology but only about "an" ontology of technology (recognizing, in fact, that there are possible many, not only one). Perhaps, the best working term would be that of "an ontological model", again, without any claims of singularity. Here, I must restate the fact that the discussions about ontology of technology diverge into at least two opposite points of view. One is expressed by the idea that ontology somehow finds itself in opposition to history. No matter how curious it might seem, I believe that this idea has got along with a kind of broader theoretical preoccupation towards the role of technology in the middle of human culture and society. The other is the idea that ontology must not be seen as being "out of history" or *a priori* in any sense. From this second point of view, I appreciate that a relative advantage can be obtained, as permitting to account for technological change during the time and, thus, to enable a more analytical perspective which better sustains the ontological accounts.

Ontology of technology (at least a modern one) could hardly avoid the problem of time, which is shown by the evolution of technological systems as a para-"natural" process. Thus, they appear and evolve according to a genuine descent, being subjected to a process of selection as well as to different conditions of reproduction, whose complex criteria are still known in a small extent. Every technical object is an outcome of two complementary processes: *juxtaposition* and *integration* (Kaplan, 2008).

Kaplan writes about 3 stages of germination of technical systems:

1. A new technical ensemble is made of existing or developing technical elements, characterized by a weak degree of integration. There it creates a juxtaposition composed by different technical elements, coming from independent and potentially very old technological descents.

2. During a second period of growth and maturation, the technical object still transforms itself through juxtaposing and integration. The spreading out of the mentioned item continues, but in a certain moment the technique comes to a *turning point*, where begins a massive *self-enhancement* of the item. This acceleration leads, on one hand, to a stronger integration of the technical item, and on the other hand, to an emergence of new systems based either on the specified technical object itself or on some of its components.

3. In a third period of growth, the technical descents either succeed to stabilize themselves or, on the contrary, they can regress. It is the case that the technical item (system) reaches its maximum level of functioning (i.e. its limits) and for such reason it is replaced – sooner or later – with another one, coming from a different technological descent, better adapted to a specific environment (Kaplan, 2009, 10).

A "biological" ontology of technology emerges, therefore, as possible, since many similarities between technical objects and living beings have already been established (e.g. organized and evolutionary descents; embedded ecosystems; variable degree of integration; multi-scale selection, processes of integration and juxtaposition; exponential diffusion etc.). However, as technical selection has its peculiar features, the process of their increasing complexity is likely to be one without common measure in comparison with biological evolution. The components of technical objects may be extracted from their ensembles (systems) and joined together later in many other unexpected combinations. Moreover, the changing of technical items and systems is responsible for unanticipated change of our environment (natural, social and technological):

> Alors que l'évolution biologique reste cantonnée dans des temporalités lentes, les objets techniques sont aujourd'hui au cœur de dynamiques intenses de recombinaison, de diffusion et de sélection donnant naissance chaque jour à des nouveaux agencements, susceptibles de bouleverser profondément la structure des milieux que nous habitons. (Kaplan, 2009, 11-12)
>
> [While the biological evolution remains anchored in slow temporality, the technical objects find themselves nowadays in the middle of an intense dynamics of recombination, spreading out and selection, giving birth everyday to some new combinations, susceptible of scrambling deeply the structure of the *milieus* we are living]

This quick and unpredictable replication of machines inside a world of their own – however, connected and interdependent with ours – allows many ethical entailments (e.g. whether such an evolution is good, bad or indifferent for humans). The most known and famous position is that of human's fear of being controlled and oppressed by robots, machines, technical devices etc. – a theoretical possibility in the future, which is still unclear at present moment. Neither mankind's nor machines' evolution is sufficiently known and clarified in all its essential points, so until then the imagination can go free to improvise scenarios with SF emergent evil machines, heroes (humans and robots), salvation of the world, death and resurrection of men (eventually helped by "good" robots and so on).

From an ontological point of view, the relationship between technology (through technical items/ systems) and our body is today one of utmost interest. And this is because the body can use technical items as *extensions* in order to perform physical duties or any other kind of tasks. In informatics, technical means are nowadays used particularly for construing a colossal global network (the so-called World Wide Web) – a huge collective and impersonal memory which illustrates K. Popper's 'third world'. In fact, the use of and engagement with technical artefacts involves always an enrolment of objects (as well as subjects) into an array of different kind of networks which enhance our physical and mental capabilities up to some yet unknown magnitude.

> Technical objects extend capabilities, at least in part, by their positional enrolment in systems of use – if they are removed the capability is removed too (at least until a replacement is found)

writes Lawson (2008).

Supported by technology, the human body can change its features, dimensions, weight, velocity etc. and so on, according to the parameters of various connections established with technical items and systems. In fact, one can speak about a process of *incorporation* – insufficiently understood and studied – of these technical objects within the coordinates of our body, with the outcome of living as a permanent metamorphosis.

Thus, a "physical" and/ or "informatic" ontology of technology becomes possible. Kaplan asserts that we have always been, somehow, 'cyborgs', changing our bodies according to the needs and using external 'prosthesis' on this purpose:

> Nous avons toujours été, nous sommes donc, des "cyborgs" faisant toujours appel à des prothèses extérieures pour agir, penser, créer et nous définir. Dans l'instant, 'notre corps est ou l'action a lieu' pour reprendre l'expression de Merleau-Ponty. Il change, s'étend et se réduit, se métamorphose en permanence" (Kaplan, 2009, 13)

> [We have always been and therefore are "cyborgs" needing external prosthesis in order to act, think, create and define ourselves. In a moment, 'our body is the spot where action happens', to use the words of Merleau-Ponty. The body changes, expands and retracts itself; it is subjected to a perpetual metamorphosis.]

From a philosophical (and ontological) point of view, the multiplication of technology which allows the enhancement of our body's performances is, however, ambiguous. Such an enhancement can be (and really is) a benefit for humans when it is to speak about the need to be spared of nature's outbursts, to control the environment well enough to ensure one's living or to make scientific discoveries. But nevertheless, the mass proliferation of machines leads gradually not only to a destruction of the physical nature but also to an even greater increasing of mechanization and fragmentation in the real (and virtual) world. Far from his former position (i.e. as a philosopher of nature), the now scientist has been compelled to become a quantitative technician. The source of the new savageness relies, according to B. Nicolescu, on the explosive blend of "binary thinking" – a pure product of the mind – and a technology without any humanistic view (Nicolescu, 2003, 93).

The dynamics of technology "imprints" its coordinates onto the philosophy and ontology of technology. What I have mentioned as a new direction of exploration in these fields (i.e. the context of that specific "empirical turn", which made the philosophy of technology to become more analytical and orientated to a greater extent to specific and concrete technological developments) is an argument in favour of a "historical" and "temporal" ontology of technology. A "historical" ontology is more realistic as revealed by the study of two processes governing the evolution of technology: *autonomy* and *symbiosis*.

On one hand, the tendency towards *autonomy* means that a certain technological descent aims for integration and autonomy in a would-be closed system, where it can function without the need for human supervision or intervention. On the other hand, the tendency towards *symbiosis* means that, in order to spread itself, every technological descent must multiply the interfaces both with technical and human *milieu* inside which it develops. Obviously, these tendencies are contradicting each other but, nevertheless, they have to conciliate their goals and the point of appeasement comes when it is to satisfy the human needs of approaching them.

In this respect, the evolution of the computer interface is an example that is often given. Until proper programs and interfaces were developed for computers (as was the case during the first half of the 20th century), able to allow their manipulation even to non-specialists, the user had to be a specialist in electronics, algebra, symbolic logic, programming etc. Only beginning in the 1950s there have been new operating systems and programs created, allowing the user to handle the syntactic and symbolic interaction with the computer. At about 30 years later, through the introduction of the graphic interface, the desktop, the mouse etc., the user could, at last, "forget" the intrinsic knowledge of informatics in order to do many ordinary jobs quicker and quicker with MS-DOS, Word Perfect and, eventually, with Windows. Then, marking a milestone in the history of informatics and computer science, a new machine was born – the historical "PC" (Personal Computer).

Thus, a new ontology is possible, i.e. the "humanistic" ontology of technology, which can prove itself able to encompass the complexity of two processes: the first one *externalizes* human capabilities through techniques and technical devices; the second one *internalizes* the way these machines work, inside the human being (in both its mind and behaviour). Sooner or later, an unavoidable temptation – as well as necessity – arises for modern humans: to compare themselves somehow with machines and complex technical devices.

> Il existe donc une circularité entre le fait d'extérioriser toujours plus ses propres capacités dans des machines et celui de se définir en retour comme analogue à ces mêmes machines, comme le montre bien la prédominance d'une conception mécaniste du corps humain. (Cerqui, 2009, 20-21)
>
> [Therefore, there is circularity between the fact of externalizing human's own capabilities inside the body of a machine every time more and more and the reciprocal fact of defining the human as an *analogon* of these very machines, as very well shown by the prevailing mechanicism while conceiving the human body.]

As, through the history of technology, the mechanisms' tendency towards autonomy is compensated by its tendency towards symbiosis, a reasonable attitude would be to suggest that machines are not ready to replace humans, but rather to communicate with them by creating and developing more and more sophisticated interfaces. It is the technique's ambivalent nature, that of being both non-human and an element of humanization. The truth is that this attitude towards the relationship between the human world and the world of machines (they interfere, of course, as mentioned before) is not without consequences on philosophy and ontology. Some questions might support this: is it the same ratio or extent for externalization and internalization? Do humans feel better when they are compared to machines? Is it fair to do so? And, above all, is this comparison always a "good" one? I cannot see today a strong argument in favour of mechanicism; on the contrary, many arguments against it can be collected from all-over the human sciences, beginning with the destruction of environment and ending with the shortcomings of human beings becoming cyborg-like.

A remarkable point of view concerning the ontology of technology is expressed by what can be called '*non-essentialism*'. This point directs the very question of defining technology, avoiding the appeal to classical definitions, presuming that the reader is familiar with a variety of technologies. An example in a book of essays in the philosophy of technology (Fellows, 1995) is indicated by Lawson: thus, the contributors to the volume did not concern themselves with "the essentialist exercise of defining technology", taking for granted that the reader is familiar with a variety of technologies, such as Information Technology. Lawson writes that

> ...the focus on empirically adequate descriptions of technology and engineering practices (their meaning of an empirical turn) need not of course actually generate a general definition of 'technology'. (Lawson, 2008)

Furthermore, some attempts to provide definitions of 'technology' have been accepted as either pointless or dangerous. Nevertheless, he adds, that

> the very literature that has had most to contribute to our understanding of the social dimensions of technology, has tended to shy away from any general statements about the social dimension of technology. Thus there is consensus about the fact that technology is irreducibly social, but little precision concerning the ways in which technology in general is social or of what implications follow from different conceptions of the social. (Lawson, 2008)

We can say that the non-essentialist (sometimes, even anti-essentialist) point of view concerning technology and its possible ontology, presumably derives from the already traditional influence of Wittgenstein's philosophy (more precisely, of 'the second' Wittgenstein) inside the Anglo-Saxon cultural area, where the status of "general statements" is rather problematic. Like a human being looking in vain for its proper "nature" or "essence", it seems to us less important to search for an ideal "essence" of technique and technology, in spite of the fact that all techniques and technologies have something in common. For instance, in principle, we can accept the point that *technology* denotes

> the material conditions and consequences of those activities most essentially engaged in harnessing the intrinsic causal powers of material artefacts in order to extend human capabilities (Lawson, 2008)

but this acceptance does not spare us the effort to examine on the spot each and every technique and technology in history, to see what might be their characteristic features, similarities as well as differences, how they develop and vary with the time etc. Of course, "general statements" about the social dimension of technology are welcome, as long as they do not pretend to show themselves as infallible.

*

It seems to me that the plea for a *moderated-relativistic, historicist and non-essentialist* view upon the ontology of technology would be an acceptable conclusion, be it such an ontology that is a "biological", an "informatic" or a "humanistic" one. (I draw attention to the fact that they are not at all incompatible; on the contrary, they frequently suppose and support each other). As the importance of case studies, historical interpretations and relativistic arguments came up with a whole literature, the ontology of technology is now better prepared to shape models able to theorize and explain the complexity of the technical field, issued from its multiple and sophisticated interactions with the human being.

Ontological pluralism is, therefore, a necessity for the views of today, focusing technology. From the methodological standpoint, analytical and pragmatist approaches seem to cover pretty well the subject-matter, not excluding, of course, more general and abstract considerations up to theories about social and cultural entailments of human–machine interaction. One has to have in sight all of these when trying to account for "the" ontology of technology.

The technical ecosystem finds itself today in a strong symbiotic relationship (which gets even stronger) with the human being; it is desirable that this balance should be maintained in the future, since humans are a part of this special environment and some of us see our future as becoming parts of machines. But how far can one go on this path? The answer is far from being definite. When dealing with technical systems or technologies, one must deal also with the claims for "amelioration" of humans through implanting technical devices inside their bodies. And because there are such situations that have already occurred (in the case of therapy), the ontology of technology must also account for the issues of those who volunteered themselves to have technical implants put into their bodies with the scope of "ameliorating" (maybe even *transgress*) the so-called "limitations" of their being. The borderline is fascinating not only for SF writers but also for professors, students and researchers. It is hard to say if such a choice means rather advertising, wisdom or even insanity. I believe that a thinking focused on this would try to deal with the "ontology of 'no-more humans' technology", who's subject is to be developed in some other place.

ETHICAL CHALLENGES IN A KNOWLEDGE BASED SOCIETY

SORIN-TUDOR MAXIM

An undeniable characteristic of contemporary theoretical discourse is recovering moral thinking, the central concern of any systematic reflections on the social. This is justified by the fact that serious problems that the world faces today do not seem to find a resolution other than to address ethics.

We are faced with phenomena generating deep uncertainty and insecurity: alongside other rich countries and regions there are masses of people that starve. There are countries with political stability but also some where infighting cause thousands of victims; fundamentalist threats and exacerbated nationalism stress the spectrum of risk factors up to the limit of acceptability. Neither the economic nor the scientific and technical areas send us more encouraging signs: the pride of the organizations belonging to knowledge society to believe that free creative spirit may however extend the permitted limits of knowledge lead to concerns of finding solutions to questions that may never be asked.

Questions regarding human cloning, genetic manipulation, development of human artificial intelligence, the development of artificial intelligence to the point where it can take on not only physical but also intellectual human tasks, and not least, the intemperate interventions on nature and on life, with the most uncontrollable effects on the future of mankind as a whole.

Consequently, the importance that the ethics acquire is amplified as a guarantee of inducing moral, fair and responsible behaviour to all the participants and the social actors in order to enhance the chances of resolving crises and eliminating risk factors or at least their suspension in order to avoid the beginning of the fatal risk: the extinction of humankind.

1. An Ethical Paradox

Gilles Lipovetsky stated (1993) that faced with such a status-quo:

> A name, an ideal unites the minds and re-animates western democracies at the end of this millennium: ethics. [...] 21st century will be ethical or won't be at all.

Here's the first challenge that the knowledge based society submits to the moral thinking. We are dealing here with a paradox: raised everywhere - in business, politics, media, philosophy or justice, as bioethics, techno-ethics, environmental or ethical globalization - ethics does not enhance itself in any way. This is because, as the same author states, we do not have a moral; landmarks have disappeared, duties also disappear and we inherit the void. In other words, we are witnessing a dramatic relativism of values, and moral values are no exception. The "dusk" of the modern morality is not replaced by an ethical claim of universality, but by an ethical pluralism, the so called "minimal" or "weak", removing any imperative content.

These types of "persuasive" ethics could thus respond more adequately to the demands of the knowledge based society, as belonging to the broader scale of "post-moralist" societies. Hence it can be concluded that there is not a generic human world. People are different, with different problems. In the spirit of the ethics developed during the "post-moralist" period there is not only a solution to human problems, but multiple solutions. The acceptable compromise, tolerance, "reasonable" morality - not perfect - the science and practice of negotiation are all solutions permitted and justified in terms of such ethics. To these there can be added a purely pragmatic argument: in such a knowledge based society, where information circulates and is generalized, in which virtually everyone can know what is essential about the other, such a behaviour is even counterproductive without morality both at the individual level, and especially in communities or organizations. To build an ethical basis of these new premises no longer seems a utopian project or a hopeless one, but a realistic theoretical approach that anticipates a new reality and that aims to prepare the actual man that acts in concrete situations.

Jürgen Habermas' position developed in his famous work "On Moral Conscience" (1983) seems very significant seen from the perspective of communicational action. Speech ethics, as a result of communicational action, *does not provide content guidelines or argumentation rules* - as in the case of the ethics based on a fundamental principle of universality - but a *procedure* with many assumptions that guarantee the impartiality of making moral judgments. As such, this type of moral discourse has the essential universality feature as with any genuine ethics, not based on the fundamental principle of universality but in the sense of *inclusion of all concerned people*. This ethics corresponds to a society based on communicational action thus oriented on understanding and knowledge. A world that we think in the present but that belongs to the future.

2. Prospective Ethics

Certainly the whole culture of knowledge-based society is a culture of change, and this means that the management of the future is a management of change:

> To survive and succeed each institution should develop into an agent of change. The best way to manage change is to determine it (Drucker, 2004, 75)

But in order to cause the change they must act on mentality, behaviour, attitudes, value system, and rules. Even the ethical codes of communities and organizations need to change, even imposing a redefinition of fundamental values and ethics like the spirit of tolerance, loyalty, responsibility, solidarity and fairness. In a world of interdependence and the information revolution, fairness in any human activity is not an option, rather the only solution to survive in an area where everyone is becoming more precisely and comprehensively informed about other social participants, competitors and collaborators; it is also a moral choice for the new community that information technology dominance makes more and more open. And there is also a need to enforce and manage through education and ethical culture change.

In the classic approach, ethics makes a speech about "what is", about the moral life and spiritual practice of individuals or / and communities, so it could, on this basis, prefigure the ideal world of "what should be".

Extremely high rates of change, representative of the information society based on knowledge, require that ethics assume a new function: normative and persuasive. As an expression of systematic moral experience of humanity, ethics must exercise a prospective vocation.

The concept of prospective ethics concerns, as a new moral dimension of discourse theory, to not be content only with the reflection of reality but to nod at the understanding and description of a world that isn't yet, but could be, as it doesn't contradict current trends and social development. The importance of this prospective function is found in the ability of the ethical discourse to anticipate future social reality, with the good and the bad, with mutations more or less significant in the moral behaviour of the individual or group. The fact is all the more worthy to be considered, if we consider that the morals and attitudes are the results of education or, the social actor of the future is the product of present education. It takes time, but fast paced changes can lead to an individual that is always out of sync with the standards of the period if one does not give concern to creating an adequate personality model.

What impact does globalization have? What impact will the new information technologies have on individuals and communities and the human condition in general? How can the excessively fast pace of change be managed? How to prepare people to withstand a reality where "normality" is change itself, stability only representing "the accident"? All of these are questions that prospective ethics must find answers to in order prepare humanity for the challenges ahead.

3. Ethics Face to Face with the Challenges of Technology

There is a significant difference between the changes introduced in the history and civilization of mankind by the industrial revolution and what we are living today as a new revolution; a starting point for a knowledge based society.

In the first case, the gains of the art and science

> led to *economo-centrism*, i.e. the idea that religion, culture and others have a secondary significance and, according to Marx, are determined by economics. (Toffler & Toffler, 2006, 357)

In the second case, the social wealth

> is based more and more on knowledge and sets economy in its place as part of a larger system in which issues such as cultural identity, religion and morality are *back into the spotlight.* (Toffler & Toffler, 2006, 357)

In both cases the role of technology is crucial, but repositioning the economic role shows that today, while always bearing the mark of technology, especially because of its spectacular development, they are much deeper and not just of an economic nature or achievement on a material level: objects, artefacts, productive processes etc.

A first consequence of this is social awareness of the need to always submit the progress of scientific knowledge and increasing power of the technique to systematic moral judgments; bioethics, business ethics, environmental ethics, techno-ethics occurred just because of this requirement.

Contemporary ethical debates insist that, through technology, man transforms the nature of human beings and things, creating his own human world; the world of culture and civilization. Unfortunately, many technologies have produced risky usages - genetic manipulation, weapons of mass destruction, information control by interfering in the privacy of individuals - or negative effects - such as pollution of the planet or drastic reduction of natural resources - or early applications, without a analysis of effects and their long-term consequences. Then, some achievements of technology are put into the service of at least some disquieting, if not morally questionable, issues such as cloning techniques, euthanasia, abortion or even intemperate drug interventions with unpredictable and uncontrollable mutations on the human being.

On the other hand, a positive consequence is that in a knowledge based society, in a world of inter-comprehension and communication, which is repositioning the ethical foundation of the whole social action, it is increasingly difficult to invoke the argument of the predictive inability or lack of intention in defence of the undesirable or even disastrous consequences of irresponsible use of technology.

A third consequence that today's technology has on ethics is linked to the fact that values cannot be based solely on what is or is not desirable for one side or the other of a whole, while ignoring the stability or smooth functioning of the whole itself. Thus, JB Callicott writes:

> At the level of social organization, society's interests may not always coincide with the sum of their interests. Discipline, sacrifice, individual constraints are often necessary to maintain the social sphere, as a body, social integrity. Nonetheless, a society is particularly vulnerable as the danger of disintegration appears - as family members are entirely concerned about their particular interests and ignoring the distinct and independent interests of the community as a whole. (Callicott, 1980, 323)

Clearly, some uses of technology can and actually do promote particular areas, for example, accelerating the process of maximizing economic profit, or the pharmaceutical industry or the arms industry, while others can optimize the system and information administration organization. *None of them prove their value* unless they achieve the good of the community as a whole, while addressing other considerations: environmental ones, enhancement of social welfare, ensuring an area of comprehension and collaboration, with beneficial effects for individuals and community development.

In this sense, ethics is able to mobilize, to concentrate human resources and ensure cohesion in all around the value of common Good, being mediation between different and sometimes divergent interests and aspirations.

Finally the technical process revolution and the development of technologies bring problems that are far from being technical, literally, but also wider social consequences, at the level of the individual and the collective mentalities, involving political solutions which, in turn, claim reorganizing the world values. In this context, ethics is an appropriate and complex response given to democratic societies

> in order to adjust and self-adjust their course in managing social and economic conflicts by regulating contradictions in organizational management and bring harmonization of welfare interests with the moral self-assessment and appreciation. (Morar, 2004, 83)

The ethical criterion received a new content, less "disinterested" or "pure" and more related to efficiency and public interest and, through these, public Good.

Engineering, especially the extraordinary rates of renewal of the technologies triggered by information revolution, gave birth and consciousness to the "irreducible novelty" that will characterize the world of tomorrow. Consequently, on the social scale there can be identified the possibility of a future redesign from the understanding and interpretation of new social processes caused by development of technology, processes that give rise to a new culture: the culture of change.

The society dominated by the culture of change necessarily determines a high degree of uncertainty, complexity, and conflict, in as far as the assignments and social positions of participants are always different. Uncertainty, stress, and exhaustion come with social "actors" because they no longer meet the new challenges, or because the economic, but also the social, expectations that are imposed are always different. That is why the moral dimension of the personality of all those involved in the act of change gains importance creating that space of interpersonal communication and understanding based on valuing solidarity and cooperation to promote "best practices" and common interest in social action.

WHAT IS A TECHNOLOGICAL MENTALITY?

BOGDAN POPOVENIUC

Technology is a man-made reality, one which easily enslaves him to its development strategies. Modern man was already enraptured by technology; not only in a physical sense, but also cognitive and spiritual. It became the brave new world religion. Technological rationality is seen as the royal path to most (or the only) sane solutions. Any other solution for humanity's problems are heavily questioned about their secret intentions, hidden motivation, and their purported aims, but technological ones seldom undergo such scrutiny, just because they are techno-logical. Technology is not just the external material system or the process by which human beings fashion tools and machines in order to change, manipulate, and control their environment.

> Technical progress is the pursuit of effectiveness in producing objects of a given kind; (Skolimowski, 1983, 44)

hence, techniques is definitely much more than its physical products or expressions, techniques and skills.

> Technologies are bodies of skills, knowledge, and procedures for making and doing useful things. They are techniques, means of accomplishing recognized purposes. (Merrill, 1968, 576-77)

Technology, despite its common image as fancy metallic devices and high-tech products, is actually formed by the solutions, means and tools for solving the problems posed by humane existence.

> In our technological society, technique is the *totality of methods rationally arrived at having absolute efficiency* (for a given stage of development) in *every* field of human activity. (Ellul, 1964)

Its persuasive and pervasive characteristics stem from its rational origin. It ordinates, organizes, enframes and transforms reality in a regulated, predictable and manoeuvred cosmos.

Therefore, the driving force behind technology is not to be found within the devices or in the various human practices related with modern modes of production, but in its specific "noology" which systematizes and frames the mind and world. It "frames the mind" in order to dismantle the world as standing reserves for its various way of production. Everything is de-contextualized and reduced to a form of raw material regardless of whether it is land or a human population.

> Enframing means the gathering together of that setting-upon which sets upon man, i.e., challenges him forth, to reveal the real, in the mode of ordering, as standing-reserve. Enframing means that way of revealing which holds sway in the essence of modern technology and which itself is nothing technological. (Heidegger,1954, 258)

It is assumed that this "enframing mentality" is the consequence of human drive for a "precise" and "scientific" knowledge of the world. However, in the philosophy of science there is an intense debate on the relation between science and technology: if technological practice was the cradle of modern scientific thinking or scientific theory forms the base of modern technology.

Regardless of their beginnings, nowadays they are in a mutually boosting relationship. They support and reinforce each other. On the one side, the intrinsic limitation of any scientific enterprise is exponentially mirrored in the technological approach. As Romanian philosopher Lucian Blaga shows, any problem within the situation of a conscious being presents two components. First, are the external data which are exterior to the posed problem and forms the "area" of the problem. But, the decisive factor which determines identification, description and solving, is the problem's "interior horizon":

> an ideate content, that will guide us in the process of posing the problem, and that will therefore determine, to a greater or lesser extent, the very content of the solution that will be given to the problem…the content that we don't find among the objective data of the problem, and that prefigures up to a point the very response… (Blaga, 1947, 101)

In science, the area of a problem is very specific and the interior horizon is very detailed. In contrast, a philosophical problem has the whole world as its area and the inner horizon is rather broad or vaguely defined. (Blaga 1947, 104) Almost any solution to a particular problem is possible as a working hypothesis which will be later confirmed or rejected based upon its fruitfulness. Even though both ways of problem solving anticipated the solution, in science the solution is anticipated as a matter of course, while in philosophy it is anticipated as little as possible.

> To think in terms specific for a given discipline is to think in those terms that (a) determine the lines of investigation within this discipline; (b) account for the historical development of this discipline; (c) explain the recent growth of the discipline. (Skolimowski, 1983, 48)

This fixed and programmed way of development, from the posing of the problem to the solution, represents the technology of a given science. Translating these scientific settings in technological undertaking reveals that the major limitation of present technological development is a structural one. Objective exterior data represent the area, or the domain, where the technology is supposed to be applied. The interior horizon of modern technology shares the same feature as its scientific origin. The internal horizon of the problem of technology is a specific type of order concretely and overtly imposed on nature which is reduce to technological resources and its domain of action.

On the other side, the reverse effect of technological development on the modern natural scientific endeavour is reflected in the blind artificial technological development of knowledge within it. Scientific knowledge became technological. It is oriented to a narrow problem-solving dissociated from other considerations. Even the fundamental way of inquiring is limited to gaining technical know-how knowledge, while human knowledge comprises more besides this.

> What we have at present is academic inquiry devoted primarily to acquiring knowledge and technical know-how dissociated from any intellectually more fundamental concern to help us resolve our conflicts and problems of living in more cooperatively rational ways – dissociated, that is, from the pursuit of wisdom – a recipe for disaster. (Maxwell, 2007, 5)

Scientific knowledge is by definition specialized in the realm of epistemological inquiry, but its technological (methodological) determinations dissociate the human knowledge in disparate domains of activity from within, and split the human social and cultural universe in conceptual and epistemological autistic monads. Technological rationality of modern (natural) sciences jeopardizes the possible evolution of human understanding to an integral self-understanding.

> Understanding is not concerned with grasping a fact but with apprehending a possibility of being. (Ricoeur, 1984, 56)

Thus, the presumed aim of human evolution - the accomplishment of a living conscious form - is totally compromised. (see Popoveniuc, 2008a)

Every branch of technology has its specific model of efficiency and employs particular patterns of thinking for increasing it. "Thinking, specific for the civil engineer, is in terms of durability", for the mechanical engineer "is in terms of efficiency", and "to think geodetically is to think in terms of accuracy", but,

> architectural thinking is simultaneous thinking in terms of durability and aesthetics and utility, and the two latter categories are perhaps more important than the first one. (Skolimowski, 1983, 47-48)

This "engineering mentality" spreads all over institutionalized knowledge. It can be found in all researchers and academic areas, natural, social and humanist altogether, infusing and guiding the whole universe of human inquiry. In humanistic undertakings there are already detailed techniques to compose belletrist texts, create essays, make exegetic texts, and technical writing. The entire aesthetic - if it wasn't removed by the technology of musical composition or paintings – was definitely transformed into a machine aesthetic. Thinking in philosophical inquiry, specifically for the postmodernist, is in terms of deconstruction. For a philosopher of science, modern technology has to be grounded on scientific laws (e.g. Mario Bunge). For a sociologist, technology captures the entire episteme of the author (e.g. Harvey Brooks). For him technology must have not only a sociological dimension, but it must be sociological in its nature. If there is to be such an entity, as an object of study, it must have or apply to a "sociologically significant scale". Every theory, every "-ism", is or at least implies a mental technology of organizing available data from its domain according with its core structure of conceptual analysis.

> Specific branches of learning originate and condition specific modes of thinking, develop and adhere to categories through [which] they can best express their content and by means of which they can further progress. (Skolimowski 1983, 46)

On a regular basis technology is seen mainly through tangible, material forms as "hard technology", and less as "soft technology" involved in human psychology and behaviour or the ways of performing something. (Jin, 2005, 113) From this perspective the "hard technology" is equal to concrete devices (tools, instruments, equipments, machines, plants, industrial processes etc.) and "soft technology" to knowledge capital (knowledge, information, expertise, skills in organization, management, marketing, collections of ideas expressing goal, functions and rationality of the efforts i.e. a fusion of knowledge, procedures and rationality). Because of the prevailing impersonal, device-related and substantivistic conception over technology which focuses on what technology *means*, and not on what it *does*, and *who*, *why* and *how* it is made, the profound socio-cultural aspect of technology is usually is concealed. A constructivist approach, as the one made by Andrew Feenberg, reveals instead two phases of technological process: a primary one of functionalization related to a subject's program and a secondary reflexive one of practical realization, which transforms and reshapes both technological object (objectification) and subject (subjectivation). The primary instrumentalization (functionalization) objectifies the nature and puts it in use by isolating and taking objects out from their meaningful specific context (de-contextualization), de-worlded things being stripped of technically useless qualities and revealed as resources (reduction). The subject is isolated from the effects of its actions on objects; technological settings being designed to create autonomous decision making systems (autonomization), the huge impact on nature turns into a pure functional relation in which the subject places itself in a very advantageous situation to use natural law or technology, enhancing the object proprieties (positioning). The reflexive meta-technical practice, or secondary instrumentalization (realization), combines and re-embeds decontextualized technical objects in the natural environment (systematization) through ethical and aesthetic mediating processes which supplies objects with secondary qualities that embed them in their new social contexts (mediation). In this second part of the instrumentalization process subject's acts, as deeply embedded as object, add up or match to a craft, i.e. the reverse act of tools on their user (vocation). The strategic control of workers and consumers through positioning is compensated for by the existence of a certain "margin of manoeuvre" (initiative) which belongs to the subordinate position in capitalist hierarchal society that can support conscious cooperation in the coordination of effort and user appropriation of devices and systems. They contain an opportunity to change, individuals having a flexible area of manoeuvrability within constraints. (Feenberg, 1999, 202-210)

In this perspective we can better understand the technological driving pattern of many conducts and actions from modern day professional activities. For example, psychoanalysis (as therapeutic practice) has its own technology of *de-contextualizing*. It *reduces* the manifest behaviours in symptoms of passion or libidinal urges, and the psychoanalytic object - the client – has his behaviour restored to the societal standards *systematized* through complex therapies and procedures which has a ritualized and ceremonial *ethical* and *aesthetic mediation.* The psychotherapist or theoretician is isolated from effects of its mental actions through a sort of mental shield (the psychoanalytical findings doesn't somehow apply to him). This *autonomization* of psychoanalytical technology is amplified in professional actions, by very strict procedural steps assured by regulations, and endorsed by its professional guild. The psychoanalyst has also a very advantageous *position* related with the client (expert) and somehow his perspective does not touch upon itself relating to the awful findings on human nature and psychology entailed by its theory. It always applies to the other. The various therapeutically practices have a *systematic* effect and work together to combine diverse behaviour and conduct regulations in a unitary fully-functioning personality. At the same time as the psychoanalyst with its prerequisite *vocation* of counselling have also a larger or smaller liberty of moving in his therapeutic relation (*initiative*).

In these conditions the entire human episteme seems to look like a huge *meme-plex.* The technological development of knowledge is the utmost illustration of Dawkins thesis on evolution of life on earth, as the perpetuation of self-replicating units of culture i.e. memes. Technological development of scientific knowledge and its technology is epiphylogenetic extension of selfish genes' evolution through technological replicators (or technomemes). Consequently, we risk becoming subordinated to the propagation of memes, robotically and at any opportunity, as meme-bots. The technological thinking turns all human actions, whether conscious or not, towards the complex interplay "between memes, genes and all their products, in complicated environments." The human self ceases "to be the initiator of actions, it does not have consciousness, and it does not 'do' the deliberating."

> The self that is supposed to have free will is just a story that forms part of a vast memeplex, and a false story at that. (Blackmore, 1999, 237)

Does that mean there is no escape, as Susan Blackmore decrees?
Or perhaps, although,

> we are built as gene machines and cultured as *meme machines*, [we still] have the power to turn against our creators. We, alone on earth, can rebel against the tyranny of the selfish replicators. (Dawkins, 1989, 201)

Jacques Ellul considers there are "five conditions necessary that an opening on the technical problem can even become a possibility": the diagnosis of technology as a problem and awakening of men to the danger, demolition of the devastating myth of technology to make visible the common place that it really is; changing the perception of technical objects and constructing a detached attitude toward them; elevating a live philosophy solely able to defeat the overpowering ideology of technology, and, the almost superhuman task, engaging in a dialog with technicians. (Ellul,1983, 97) Our only salvation from total technological determinism, from the complete surrender to the intrinsic reasoning which purports the autonomous evolution of technological systems, in order to avoid being caught in eutaxiology of technology and to be hence in a subordinate position to technological development, is awareness. Mental technologies and technological development have evolved beyond the modern human consciousness and awareness. The technological application and systems are out of reach for the present comprehensive human level (many technologies and applications can "think" and "respond" and "perform" more smartly than the smartest humans).

> The world has achieved brilliance without wisdom, power without conscience. Ours is a world of nuclear giants and ethical infants

said the general Omar N. Bradley. We have to bridge our individual and planetary science with our cultural consciousness; our rationality should meet (self) awareness. For this task, cultivation among the scholars is necessary, and research from both natural and social sciences, and not only that, but of epistemological *subjectivity awareness* with its two components: *ontological positioning awareness* and *self-inclusion awareness*. It is meant to counterweight the blind automatic development of human knowledge, driven only by the purpose of technological eutaxiology expansion.

This research awareness means to move beyond specialized inquiry and having always in mind the particular phylo- and epigenesis of any human understanding. The technological part of human intelligence, the faculty to create artificial objects, in particular tools to make tools, and to indefinitely vary its makings, where theory is considered one of the most powerful kinds of tool ever made, was overbid. The Mumford contention, that conceiving man as a tool-making animal obscures the truth that "man is pre-eminently a mind-using, symbol making, and self-mastering animal", is futile as long as modern technology is not merely science and not even applied sciences, but more "a form of human knowledge." (Skolimowski, 1983, 43) Understanding could be, and maybe already is, as technical and unaware as any other non-living material device. The ideate tools of modern science have the same functions, serve the same purpose and have concrete effects like the knapped flint, chopper, scrapers or pounders. The phylogenetic development of biological cognition culminates in the rupture brought by the exteriorization of the inorganic trough technology. The latter is the prolongation of biological development through

> epiphylogenesis of man, meaning the conservation, accumulation, and sedimentation of successive epigeneses, mutually articulated. Epiphylogenesis is a break with pure life, in that in the latter, epigenesis is precisely what is not conserved (…) Epiphylogenesis bestows its identity upon the human individual: the accents of his speech, the style of his approach, the forces of his gesture, the unity of his world. (Stiegler, 1998, 140)

More simply put: the human is that living being who can pass on its experience not merely through its genes, but through the conservation and dissemination of cultural knowledge. But the intelligence is the flexible capacity to choose and invent, and even if it is less genetically predetermined than the instinct, it is not completely independent by its organic substratum. And, as genetic evolution demands, the original function of knowledge (cognitive information) is that the survival and reproduction of those organisms that use is it more likely. Therefore the organisms with better knowledge of their environments would have a higher rate of surviving and reproduction. Hence, we have a phylogenetical evolution of knowledge which depends on the degree to which it helps its carrier to prevail in natural selection through its environment. Given the three-folding complex anthropological environment (natural, social and cultural), intelligence is not only species-related, but also socio-ethnic and cultural-informational. Thus, human knowledge is prone to three underlying interest categories: towards technical control, practical interaction and emancipator ideals; that is, techno-centric, anthropomorphic, and ideal. (Habermas, 1972, 314)

Technological mentality exercises only the instrumental part of the last one of the above categories. It transforms the humans into informational producing machines, or better, in some being designed for secret objective scientific knowledge meant to support technological development. Although the *objective knowledge*, as systematic and predictable organization of the universe is possible, the experience of *knowledge* remains

> something personal and private that cannot be transferred, and that which one believes to be transferable, objective knowledge, must always be

> created by the listener: the listener understands, and *objective knowledge* appears transferred, only if he is prepared to understand. (Maturana, 1970, 5)

Hence, any "objective" rational system depends on the individual positioning into the tri-sided domain of interaction circumscribed by its tri-unitary nature: bio-socio-cultural. Man can't step outside the limits impose by his domain of discourse. The position relativity of any understanding requires being aware about all three dependences: biological, social and cultural. In the first place, man's capacity to know that he can rely on his biological integrity because cognition is a biological phenomenon as well. Cognitive technology acquired through evolution is one of the strongest bestowed on human beings, their material (artefacts) products (technological instruments) enlarge our cognitive domain and generate new modes of interactions. Second, any cognition depends on language, because it is socially oriented mostly by linguistic behaviour, hence it is ultimately relative to the social settings within it has happened to develop.

Third,

> man is a rational animal that constructs his rational systems as all rational systems are constructed, that is, based on arbitrarily accepted truths (premises); being himself a relativistic self-referring deterministic system this cannot be otherwise. (Maturana, 1970, 57)

Hence, the first element of the required epistemological awareness demands that any researcher keep in mind two things. First there can be no individual objective knowledge uninfluenced by the subject's ontological position in the universe. And second, that the production of objective knowledge is rather detrimental when it is no longer related to the growth and enhancement of life on Earth and in the Universe. As Technics, the technological scientific knowledge came to be "formed by an accumulation of means which have established primacy over ends," (Ellul 1983, 86) and hence its development is unpredictable and ambivalent, for every problem solved, another (or many others) could arise. Technologized scientific research forsakes its ultimate goal of life evolution - the accomplishment of universal self-consciousness. (Popoveniuc, 2008a)

The other component of epistemological awareness is the necessity to become aware of the ultimate propriety *of self-reference of any comprehensive theory*. Man is a deterministic and relativistic self-referring autonomous system whose life acquires its special characteristics through self-consciousness.

> For the observer an entity is an entity when he can describe it. To describe is to enumerate the actual or potential interactions and relations of the described entity. Accordingly, the observer can describe an entity only if there is at least one other entity from which he can distinguish it and with which he can observe it to interact or relate. This second entity that serves as a reference for the description can be any entity, but *the ultimate reference for any description is the observer himself.*[*] (Maturana, 1970, 8)

In the sciences, as in knowledge in general, any description of a phenomenon, social and human, comprise volens nolens their individual author, although, he seems to believe he is in illusory objective stance, positioning himself in a privileged outside-this-world frame of reference. But even the maturation of natural sciences, in its endeavour to achieve a Grand Unified Theory (GUT), was forced to integrate the masterpiece - quantum physics - that achievement that the human consciousness, under the instance of the observer, is part of the description or model of the system. In other words, the scientific meanings cannot transcend the particularities of *how*, *when*, and, especially, *by whom* something is observed, described and understood. The cognitive technologies (which form the infra-structure of any scientific theory) re-shape and reorder the reality accordingly.

Technological mentality and the required self-awareness could be noticed within the act of conceiving the current paper, as well. Like other professional scholars I am trained and gained experience in delineating a given topic, gathering data, using various criteria for sorting, comparing and filtering them, designing, reasoning, and making, based on available data, a comprehensive description. All these are based on my personal skills which I had to develop by the regulations and requirements imposed by my situation of cerebral being living in academic settings. The technological mentality in this case consists in variation (combination, mutation, recombination), replication and hence, replication of dominant conceptions on this topics. I and my mind am just the vehicle of self-explanatory and self-grounded development of technology through its accompanying theory about it.

In Dawkins' *Memetics* terms we'll say that all actions of making this paper pertain to technological tools developed by the main thesis of the paper for surviving in the competition with concurrent ideas which developed and apply analogous technologies for their own prevailing. In fact, this seems to be the essence of science,

---

[*] *Italics added.*

> fundamentally a process; a set of methods for trying to distinguish true memes from false ones. At its heart lies the idea of building theories about the world and testing them, rather like perceptual systems do. (Blackmore, 1999, 202)

The human knowledge is only a worldwide ideo-sphere of understanding emanating from human beings' consciousness and fixated through exogenous memory's technology. It's up to us to not let our consciousness slip and fall under the control of its own creation.

IS THERE A CULTURE OF ENGINEERING?

EMILIA GULICIUC

The current structure of culture, strongly focused on science, on knowledge, and on their practical applications, must be nowadays associated with a form of adequate reflection, thus *a philosophy*, as specified by Laura Pană (2001), one of the Romanian authors interested in the problem of involving philosophy in the deep aspects of what is called engineering.

Our primary meditation theme is the meaning of the philosophical utility of engineering. Hence, we must start with questions like *what is philosophy*, continuing with *what is engineering* and with the need for a direct relationship between philosophy and engineering, based on the finding that today (more precisely, until today) these two aspects of culture do not seem to have had any connection.

In order to answer the question *what is philosophy today,* we must find, paradoxically, the conditions of its occurrence in the ancient Greek society that gave birth to it; at least to the name of *philosophy*. In this line of argumentation, the ideas of Aurel Codoban are very helpful. Philosophy emerged in a context of a crisis regarding values and traditions, when the Greek man had hesitations when deciding his own existence and his own behaviour, in a context when religion ceased to have the power to rigorously prescribe attitudes, norms and behaviours. Thus, philosophy emerged when man took his own existence in his hands, says Codoban.

Sharing knowledge with science, philosophy has often been considered a science. In the fully modern form of German philosophy, philosophy has even been considered *the absolute science of absolute* and in a more moderate form, *a science of all existence* and of the general laws that are driving it. Science, the modern scientific knowledge at least, has instead sought to further delimit as precisely as possible its object of study.

These different ways of defining, have produced the ironic assertion that *if the scientist is the man who knows almost everything about almost nothing, then, the philosopher is the man who knows almost nothing about almost everything.*

Today this difference, theorized for many centuries, proves to be superficial, as science claims totalizing truths, while philosophy, after Hegel, has renounced building totalizing systems. At the same time, science claims that it is the only one in control of the truth, since she can verify and test it. Compared to current scientific knowledge, philosophy can be considered as knowledge only if fields like astrology or alchemy are considered knowledge as well, mentions Codoban Aurel.

This resulted in a reorientation of contemporary philosophy, not towards the problems of knowledge, not even to those of existence, but to *language analysis*. Alienating philosophy from God (committed in antiquity), from man, from knowledge, and from the practical issues (in the contemporary era), marked the emergence of a strong crisis of "meaning" in any form of culture, of civilization and, after all, of existence.

The characteristic of ancient philosophy *to say something meaningful about man and the world* becomes a goal of the contemporary world. But its role should not be limited only to being an existential hermeneutics, an interpretation of existence, of the world and of man, even though it is an erudite one. It must provide a profound understanding of the world and of the individual, understanding that can alter the existence of the one that touches it. In its essence, philosophy should be *an interpretation of the meaning of human existence and of its values, but a changing interpretation.* It changes it, not only by reducing it, not by dropping it, but also by enriching and extending it.

We thus reach to the primordial sense of philosophy, but radically go beyond it. The Greek philosophy was not a *direct interpretation* of the world, as universally believed, but a cultural discourse applied to other cultural discourses, *an interpretation of mythological literature* (especially the Homeric poems, which are in the centre of the Greek educational system), was not a discourse of knowledge, but a discourse of *signification that occurred in the transition from sacred experience, to everyday profane experience.*

To be able to aspire to its old time prestige, philosophy must pass to a direct (not mediated) relationship with the world, exactly at the opposite side from where it stands today: *an interpretation of the meaning of its own discourse.* In our opinion, philosophy should follow this direction, in order to come together with science and especially with engineering.

We could add another perspective. Centuries ago, philosophy was regarded as love of wisdom. This position leads us nowadays to an even bigger puzzle, as we must find the answers of other questions like: *What do we mean by wisdom?* Why *love of wisdom* and *not theory of wisdom?* How can we learn or teach *the love of something*? What link can exist between engineering, technology, science and wisdom? Looking for answers to these questions in dictionaries and specialized books, the result is unsatisfactory. Thus, one question is born: *what use is philosophy to areas that seem so distant from it, such as engineering and, even more, such as wisdom*?

Moreover, mankind has never been attracted by wisdom; otherwise we would not be able to explain the fact that philosophy (as the love of wisdom) was only *the privilege of an elite*. The philosopher, the wisdom seeker, could not answer a question that persisted along the time: *that is the subject of wisdom.* The contemporary essays did not clarified the terminology, nor did they bring anything in addition to the allegations of Plato's dialogue Charmides, which ends with the pessimistic ascertainment that, in fact, wisdom[1] is not useful for us, nor can we know if it corresponds to any reality. All attempts to define wisdom[2] and its subject are only tinted repetitions of those issued by Plato.

The conclusion is simple: if we seek philosophy only to shelter under her wisdom umbrella, we risk being deprived of the most basic shelter and comfort. Philosophy can survive, in itself, only if it goes beyond the intent of interpreting the world, of purchasing a harmonious knowledge, towards a set of forms and levels of *action.* At this point, communication between engineering and philosophy becomes not only possible, but also necessary. After all, the changes in the contemporary world determined by the technical, engineering and scientific creation, need a philosophy cut out for them.

And so we reach the second aspect of our endeavour: until now we have seen what makes philosophy meet technology, now we will analyze what engineering may present as interesting points for philosophy. The definitions regarding what *engineering* represents are not so many and contradictory as the ones for philosophy.

Even if the term *engineering* was not used in the early days, engineering activity, as a line of work and creation, is present even from

---

[1] Wisdom as a way of doing things calmly, wisdom as a kind of shyness, wisdom as a way to pursue your own ideas, wisdom as knowledge by itself, as the wisdom to know that you know and to not know what you do not know, or in conclusion, wisdom as the knowledge of good and evil, as Mirea G.A. suggested in his blog.

[2] *DEX* sustains that the terms *wisdom* (the Romanian word/term searched is: "înțeleciune") comes from the Latin *intellectio, onís* and has two meanings: 1. *higher capacity of understanding things;* 2. *Thrift, prudence, temperance, determined by experience, spirit of prevision.* On: NODEX we will also find the assignation: *the faculty to judge and deeply understand things.*

the first stages of human civilization. At the same time you can observe that the expansion and acknowledgement of engineering activities and engineering as a profession, has occurred together with the technical-industrial revolution of the 18th century. Since then, you can see another aspect; that of the depth of interests and domain specializations, which, as science and technology evolved, they have emphasized. From the *universal engineer* and the *artist* defined through activity and his creation, we have reached the *research engineer* and the *specialist* in every line of human activity.

Engineering is defined today *as a branch of applied sciences, which follow the use of natural and human resources for the benefit of mankind, with main interests in the inevitable limitations, in proportion with the requirements implied by these resources.*

Unlike scientists, who study nature and natural phenomena, in order to state principles, axioms and theorems, engineers *apply* mathematical and physics principles to *create* a *concrete* product.

In the context of contemporary society, this distinction vanishes. The collaboration between the engineer –who assimilates more and more scientific knowledge – and the scientist has intensified, and the public treatment awarded to the two categories tends to equalize.

Also, it has become more obvious that the tendency of growth of the collaboration between the fundamental research, applied research and production, is erasing the boundaries between the research categories, increasing the degree of interdisciplinary, and assimilating the intrepid behaviour of the research teams.

The content of the engineering profession can be synthesized as an anticipation activity for the development and evolution of a system and programming interventions, in order to achieve and maintain its stability.

The anticipatory character includes engineering in the category of conception activities, by way of the fact that the mental image precedes the concept, unlike the execution activities in which the concrete representation or the mental image of a pre-existing object is the basis of its reproduction. Engineering refers to the elaboration and *achievement of the means used* to practice the most different professions, in different human activities, from the least to the highly qualified, which gives engineering a technical character.

Therefore, the activity of engineering can be characterized by the criteria of the novelty degree, on the size of the activity product, and also on the methods used to create that product.

The products of engineering activity answer social needs, immediate or perspective.

The history of scientific discoveries has lead to the acknowledgement that, in most cases, ages create their innovators which by their interests, respond to social expectations, depending on the general culture level, the development extent under informational and instrumental aspect and, not least, the attitude of the age. Under these conditions, when society is ready to accept the new idea, understand it, and use it properly, its assimilation and implementation is fast.

We can say that the integration tendencies of the inventor have been directed to the integration tendencies of the society. But often, we can see also the creator's insight for the non-acute tendencies in the social plan, discoveries which will focus the interest of mankind. To support this, we can recall the flying machine of Leonardo da Vinci, before which, the idea of humans moving in the air could not exceed the myth of Icarus, and which needed centuries to become real, in the social plan.

The degree of pertinence between the *social need and the utility of the engineering products*, gives social value to these products. Consequently, engineering is a conception activity in the technical domain, which assumes a creational psychic instrumentation to create new products, valuable for the social aspect.

Engineering is deeply anchored in reality, but often it is too caught in it, so it can't obtain the necessary distance for an overview perspective. The world is complexly structured, open, dynamic, far from balanced, and characterized by continuous development and renewal. It can't be known but by using interpretations in which the imaginary combines with reality, science and practical actions with philosophy, or with art, or with human-social sciences.

Where is philosophy localized in this complex mechanism of human evolution and what can be its relation with engineering?

Just like an entity is based on something different that itself (on nonentity), like the body is based on something else (the soul), in this same manner philosophy and engineering should support one another, like something different, but indispensable, because the opposite of philosophy is neither science (as philosophy has been considered, sometimes a science, or *the science*), nor religion, which is its origin and also shares a lot of common aspects, but even better, it is the *practical action*, *applied execution*, which engineering assumes, by definition. In the same way, engineering is not fundamentally different from science, (if we think of the first meaning of *tehne*), but, the element which could establish it, by the fact that it's totally different, it's philosophy. The theory of opposite attraction should work also in this case.

On the other side, even in the first half of the 20th century, philosophers have started to study the structure of the technical culture, the technical values, and the technical time.

The tendencies and features of the current age, determined by the development and the diversification of the technical culture, generates and artificializes, without precedent, the majority of the aspects of individual and social life, imposing this aspect on philosophical reflection.

Now, at the start of the third millennium, people talk about the *philosophy of techno science*, the artificial technical environment and the *philosophy of the artificial*, the *philosophy of artificial intelligence*, the *philosophy of robotics*, the *philosophy of virtual reality,* and even about the *artificial philosophy*.

In this way a basis is created for a *new human ontology*.

The relation with philosophy does not stop here.

If this relation has not become operational in a long time, this is caused, on one side, by the fall of philosophy and its transformation in a *garrula philosophy*, a *blabber philosophy*, a *talkative philosophy*, which has given up to its fundamental content, as it was defined in the beginning; and on the other side, engineering, which has retained some of its attributes, exclusively the technical ones. Maybe not by chance *the engineer has been defined as a person with a technical training, both practical and theoretical*, without referring to his human dimension in any way. Most of the time, if not all the time, *the reading and interpretation of nature's elements* is not accessible to anyone. Nobody can "*discover*" the principles or the laws which define a phenomenon or a process from a certain part of creation. The discovery of laws is a very delicate intellectual process, which presumes a profound adequacy of the real in nature with those conceptual categories that philosophy can transmit in the best way.

There are other aspects for which the closeness of engineering to philosophy is equally significant in their nature. In its original form, *engineering* derives from *ingenium* (which had the initial meaning of *spirit, born ability, genius*), which has lead to *ingenuus* (at its origin, *free man*, transformed then in *simplicity*, *naturalness* bonded with *sincerity* and *naivety*, *candor*, *innocence*, *purity*, *moral cleanliness*). In this final meaning, *naivety* is their meeting place, because, we cannot overlook the acknowledgement that philosophy starts with the *astonishment* of man in the world, and this astonishment is the expression of naivety, without it nothing durable can be built.

Considerations about the Concept of Reverse-Engineering

Cătălina-Daniela Răducu
and Mihai Floroiu

Business and technology are public and social enterprises. They gather people in view of common interests, thus being created communities which share principles, values, practices, results and even ways of thinking. But such communities are not alone in the social space; they are not independent of other similar communities that exist in the same countries, continents, or at the global scale. They might collaborate for attaining similar goals or compete for the primacy in attaining the same results. One of the most basic requirements in this "drive for success" is the need to know what the competition is doing. And one of the easiest ways to obtain such knowledge is through observation of the products created by the competition; the simplest thing to do is to buy the product, analyze it and get to know the way that it has been created. By doing that, the specialists not only get to know the internal structure of the product but also the technological challenges that are to be faced in manufacturing a new, independent version of the product.

This is where reverse engineering comes in: if we take a look back into the past, reverse engineering had an important role in disseminating the results of various technological fields, from large products as aircrafts down to the smallest products such as microchips, from war technology to the electronics industry. Reverse engineering does not relate only to the need to disassemble and re-engineer old parts that have gone out of service and need to be replaced in the field of equipments such as military systems, nuclear reactors, airliners, ships etc.

Reverse engineering also, and mainly, relates to the need to make use of the representation of a system which is, to put it philosophically, a simplification of the real world; its analysis and representation provides all the necessary means for delimiting, measuring and studying the main characteristics of the system, thus helping the engineers to define, describe and configure all the structural and functional units of the system with the goal of attaining better knowledge and eventually, a better way of action in the process of creation of other systems.

It is impossible to approach the concept of reverse engineering without bringing into discussion the vital legal issue of intellectual property. In fact, the two are inseparable. The concept of intellectual property refers broadly to the creations of the human mind. Intellectual property rights protect the interests of creators by giving them property rights over their creations.

A comprehensive list of the human creations protected by intellectual property rights is to be found in the *Convention Establishing the World Intellectual Property Organization* (1967). It specifies that intellectual property refers to: literary, artistic and scientific works; performances of performing artists, phonograms, and broadcasts; inventions in all fields of human endeavor; scientific discoveries; industrial designs; trademarks, service marks, and commercial names and designations; protection against unfair competition; and

> "all other rights resulting from intellectual activity in the industrial, scientific, literary or artistic fields."[1]

Thus, it is important to underline the fact that intellectual property relates to *information* or *knowledge*, which is or can be incorporated in tangible objects which can be replicated in innumerable copies all over the world. A very important aspect to be mentioned is that the property is not in those copies but in the information or knowledge reflected in them. Another important aspect is that intellectual property rights are also characterized by certain limitations, such as limited duration in the case of copyright and patents.

A first recognition of the importance of protecting intellectual property is to be found more than a century ago, in the *Paris Convention for the Protection of Industrial Property*[2] in 1883 and the *Berne Convention for the Protection of Literary and Artistic Works*[3] in 1886. Both treaties are administered by the World Intellectual Property Organization.

Countries generally have laws to protect intellectual property for two main reasons: one is to give statutory expression to the moral and

---

[1] *Convention Establishing the World Intellectual Property Organization*, http://www.wipo.int/treaties/en/convention/trtdocs_wo029.html

[2] See *Paris Convention for the Protection of Industrial Property*, http://www.wipo.int/treaties/en/ip/paris/trtdocs_wo020.html

[3] See *Berne Convention for the Protection of Literary and Artistic Works*, http://www.wipo.int/treaties/en/ip/berne/trtdocs_wo001.html

economic rights of creators in their creations and to the rights of the public in accessing those creations. The second is to promote creativity and the dissemination and application of its results, and to encourage fair trade, which would contribute to economic and social development.

Intellectual property is usually divided into two branches, namely industrial property and copyright. While the expression *copyright* refers to the fact, recognized in most laws, that the author has certain specific rights in his creation, such as the right to prevent a distorted reproduction, which only he can exercise, whereas other rights, such as the right to make copies, can be exercised by other persons, for example, a publisher who has obtained a license to this effect from the author, the broad application of the term "industrial property" is clearly set out in the *Paris Convention for the Protection of Industrial Property* (Article 1 (3)):

> "Industrial property shall be understood in the broadest sense and shall apply not only to industry and commerce proper, but likewise to agricultural and extractive industries and to all manufactured or natural products (...)"[4].

Nowadays, industrial property takes a range of forms, the main types of which include patents to protect inventions; industrial designs, which are aesthetic creations determining the appearance of industrial products; trademarks, service marks, layout-designs of integrated circuits, commercial names and designations, as well as geographical indications, and protection against unfair competition[5].

Serious concerns related to legality have been expressed since the birth of both the concept and the process of reverse-engineering. For the defenders of reverse-engineering, things were very simple. More than a decade ago, for instance, Cem Kaner, Professor of Software Engineering, Department of Computer Sciences from Florida Institute of Technology had almost no doubt about the legality of reverse-engineering:

> "When a new machine comes to market, competing manufacturers will buy one and take it apart to see how it was built and what it does. This is reverse engineering. It is absolutely legal. It is a normal part of innovation, one of the foundations of continuous quality improvement within an industry." (Kaner, 1998)

---

[4] *Paris Convention for the Protection of Industrial Property*, electronic source cited.

[5] According to the World International Property Organisation study *Understanding Industrial Property, WIPO Publication No. 895(E),* http://www.wipo.int

In an attempt to define the concept, we may say that "reverse engineering" encompasses any activity that is done to determine how a product works, to learn the ideas and technology that were used in developing that product. An important specification is that, usually, this is done without any knowledge of the interior workings of the product, and reverse engineers have to guess at how the internals are implemented by observation and guesswork. This is why reverse engineering can be seen, broadly speaking, as the process of extracting know-how or knowledge from a human-made artifact.

Reverse engineering is thus the general process of analyzing a technology specifically to ascertain how it was designed or how it operates. The positive aspect of this process is that it engages individuals in a constructive learning process about the operation of systems and products. Reverse engineering, when it is used as a method, is not confined to any particular purpose; it is often an important part of the scientific method and technological development. Reverse engineering may therefore be used as a learning tool: the process of taking something apart and revealing the way in which it works is often an effective way to learn how to build a technology or, more important, make improvements to it.

According to a study published on the website of the North Carolina State University,

> "Through reverse engineering, a researcher can gather the technical data necessary for the documentation of the operation of a technology or component of a system. (...) The reverse engineering process allows researchers to understand both how a program works and also what aspects of the program contribute to its not working. Independent manufacturers can participate in a competitive market that rewards the improvements made on dominant products. (...) The creation of better designs and the interoperability of existing products often begin with reverse engineering."[6]

It is also important to mention that in the European Union, reverse engineering a program (such as computer programs) that has been legitimately bought and studying or modifying its code is legal, as long as one does it only for his/her personal use or for "educational purposes"; and as long as one does not use big parts of the code for applications he/she might sell afterwards.

---

[6] http://ethics.csc.ncsu.edu/intellectual/reverse/study.php

The Council Directive 91/250/EEC of 14 May 1991 on the legal protection of computer programs (Official Journal L 122, 17/05/1991 P. 0042 – 0046) specifies, in its Article 6 (Decompilation), that:

"1. The authorization of the rightholder shall not be required where reproduction of the code and translation of its form within the meaning of Article 4 (a) and (b) are indispensable to obtain the information necessary to achieve the interoperability of an independently created computer program with other programs, provided that the following conditions are met:
(a) these acts are performed by the licensee or by another person having a right to use a copy of a program, or on their behalf by a person authorized to do so;
(b) the information necessary to achieve interoperability has not previously been readily available to the persons referred to in subparagraph (a); and
(c) these acts are confined to the parts of the original program which are necessary to achieve interoperability.
2. The provisions of paragraph 1 shall not permit the information obtained through its application:
(a) to be used for goals other than to achieve the interoperability of the independently created computer program;
(b) to be given to others, except when necessary for the interoperability of the independently created computer program; or
(c) to be used for the development, production or marketing of a computer program substantially similar in its expression, or for any other act which infringes copyright."[7]

The Directive specifies also that
Art. 6, paragraph 3:

"In accordance with the provisions of the Berne Convention for the protection of Literary and Artistic Works, the provisions of this Article may not be interpreted in such a way as to allow its application to be used in a manner which unreasonably prejudices the right holder's legitimate interests or conflicts with a normal exploitation of the computer program"[8].

In article 5, the Directive mentions: paragraph 3.
"The person having a right to use a copy of a computer program shall be entitled, without the authorization of the rightholder, to observe, study or

[7] The Council Directive 91/250/EEC of 14 May 1991 on the legal protection of computer programs (*Official Journal L 122 , 17/05/1991 P. 0042 – 0046),* http://eurlex.europa.eu/LexUriServ/LexUriServ.do?uri=CELEX:31991L0250:EN:HTML

[8] *Ibidem*

> test the functioning of the program in order to determine the ideas and principles which underlie any element of the program if he does so while performing any of the acts of loading, displaying, running, transmitting or storing the program which he is entitled to do."[9]

The European Directive regulates only on the field of computer programming. But other different uses of reverse engineering include: understanding how a product works more comprehensively than by merely observing it; investigating and correcting errors and limitations in existing programs; studying the design principles of a product as part of an education in engineering; making products and systems compatible so they can work together or share data; evaluating one's own product to understand its limitations; determining whether someone else has literally copied elements of one's own technology; creating documentation for the operation of a product whose manufacturer is unresponsive to customer service requests; transforming obsolete products into useful ones by adapting them to new systems and platforms. As can be seen, reverse engineering is not only used to figure out how something works, but also the ways in which it does not work, this being a very important aspect of practicing it, an aspect to which we will refer in the following lines.

There is, indisputably, a common misperception regarding reverse engineering: that it is *always* used for the sake of stealing or copying someone else's work. The law applies restrictions to the harmful practices but also gives clear specifications about the lawful ways in which reverse-engineering can be practiced. Educational and interoperability purposes – these are the ways in which reverse-engineering a program is permitted beyond any legal doubt.

Besides other similar laws being considered in Europe and elsewhere, an important act regulating the legality of reverse engineering is the Digital Millennium Copyright Act (DMCA) of 1998 which was one of the first laws to expressly make illegal certain types of reverse engineering of software, mainly for the purposes of circumventing anti-copying protection on digital movies and music. But it also affirmed the right to reverse engineer for the purpose of learning the underlying technology and for interoperability:

---

[9] *Ibidem*

> "(f) REVERSE ENGINEERING
> (1) Notwithstanding the provisions of subsection (a)(1)(A), a person who has lawfully obtained the right to use a copy of a computer program may circumvent a technological measure that effectively controls access to a particular portion of that program for the sole purpose of identifying and analyzing those elements of the program that are necessary to achieve interoperability of an independently created computer program with other programs, and that have not previously been readily available to the person engaging in the circumvention, to the extent any such acts of identification and analysis do not constitute infringement under this title.
> (2) Notwithstanding the provisions of subsections (a)(2) and (b), a person may develop and employ technological means to circumvent a technological measure, or to circumvent protection afforded by a technological measure, in order to enable the identification and analysis under paragraph (1), or for the purpose of enabling interoperability of an independently created computer program with other programs, if such means are necessary to achieve such interoperability, to the extent that doing so does not constitute infringement under this title.
> (3) The information acquired through the acts permitted under paragraph (1), and the means permitted under paragraph (2), may be made available to others if the person referred to in paragraph (1) or (2), as the case may be, provides such information or means solely for the purpose of enabling interoperability of an independently created computer program with other programs, and to the extent that doing so does not constitute infringement under this title or violate applicable law other than this section.
> (4) For purposes of this subsection, the term 'interoperability' means the ability of computer programs to exchange information and of such programs mutually to use the information which has been exchanged."[10]

In the above-mentioned article of Professor Kaner (1998), a natural question is raised: "Why do people reverse engineer?" and the author is giving an answer from the perspective of the engineer involved for many years in this process. He identifies a series of reasons that underlay their work: personal education; the need to understand and work around (or fix) limitations and defects in tools that they were using; the need to understand and work around (or fix) defects in third-party products; the necessity of making their product compatible with (able to work with) another product; the necessity of learning the principles that guided a competitor's design; the desire to determine whether another company had stolen and reused some of their company's source code; and the scientific curiosity of determining whether a product is capable of living up to its

---

[10] Digital Millennium Copyright Act, Section 1201, f (1)-(4), http://www.eff.org/IP/DMCA/hr2281_dmca_law_19981020_pl105-304.html

advertised claims that would allow them to demonstrate the falseness of these claims if such was the case.

The professional engineer`s point of view is a more flexible one; it has to do, if we were to paraphrase Basarab Nicolescu (2002), with the technology *in vivo*, rather than with the technology *in vitro*, governed by strong normative constraints.

Of course there are also ethical aspects to be considered, but lawyers and economists have endorsed reverse engineering as an appropriate way to obtain information about a product, even if the intention is to make a product that will draw customers away from the maker of the reverse-engineered product.

In their study of reverse engineering in various industrial contexts Pamela Samuelson and Suzanne Scotchmer[11] reach two general conclusions. The first one postulates that reverse engineering has generally been a competitively healthy way for second comers to get access to and discern the know-how embedded in an innovator's product. The second conclusion draws upon the distinction between the *act of reverse engineering*, which is generally performed to obtain know-how about the products of another company, and *what a reverse engineer does with the knowhow thereby obtained* (e.g., designing a competing or complementary product):

> "The act of reverse engineering rarely, if ever, has market-destructive effects and has the benefit of transferring knowledge. Harmful effects are far more likely to result from post-reverse-engineering activities (e.g., making a competing product with know-how from an innovator's product). Because of this, it may be more sensible to regulate post-reverse engineering activities than to regulate reverse engineering as such. (…) Acts of reverse engineering typically take place in private and are more difficult to detect than post-reverse-engineering activities (such as introducing competing or complementary products to the market). They are, as a consequence, less susceptible to effective regulation."[12]

In view of the analysis we have done so far in our paper, a conclusion has to be drawn. Is reverse-engineering legal or not? We could say, without exposing ourselves to the risks of error, that reverse-engineering can be characterized as an important part of the process of innovation, being likely to determine variations on the product that may lead to significant advances in the technological field in which it is practiced.

---

[11] Pamela Samuelson and Suzanne Scotchmer, *The Law and Economics of Reverse Engineering,* The Yale Law Journal (Vol. 111: 1582-1663), available at http://www.yalelawjournal.org/pdf/111-7/SamuelsonFINAL.pdf

[12] *Ibidem*, p. 1651

Sometimes the competitive reality of reverse engineering may act as a stimulus for the inventor to develop new, original ideas. We should not overlook the economical dimension of the phenomenon investigated so far, because reverse engineering may promote consumer welfare by providing consumers with a competing product at a lower price. These are, of course, important issues to consider and represent also important problems to be investigated and theorized, but for the present purposes of our paper, we think it is right to conclude by quoting the words of Pamela Samuelson and Suzanne Scotchmer:

> "Adapting intellectual property law so that it provides adequate, but not excessive, protection to innovations is a challenging task. In considering future proposals to limit reverse engineering, policymakers should find it helpful to consider the economic effects of mechanisms that have been employed in the past. Restrictions on reverse engineering ought to be imposed only if justified in terms of the specific characteristics of the industry, a specific threat to that industry, and the economic effects of the restriction."[13]

[13] *Ibidem*, p. 1663

CHAPTER THREE

PHILOSOPHICAL ENGAGEMENTS IN ENGINEERING AND ARTEFACTS

QUANTUM TECHNOLOGIES AND ETHICS IN THE CONSCIOUSNESS SOCIETY

FLORIN MUNTEANU

"Earth has not been a gift from our fathers,
but a mere borrowing from our children"
— Amerindian saying

1. The Need for a New Paradigm

"*Complexity* will be the *science* for the 21st century"
—Stephen Hawking

To analyze and characterize Complexity is a challenge posed to the human mind which must structure *a new ontological framework*, a *new set of concepts*, a new *methodology* and an *adequate experimental technique* that would all be qualitatively different from those used today. This framework, labelled as the Science of Complexity (Erdi, 2008), has been defined through the integration of last years' results concerning the non-linear approach of phenomena in nature, results obtained by a series of new disciplines such as Synergetics, Complexity Physics, and the Catastrophes Theory. Additionally, the appearance and progress in new related fields like Fractals, the Theory of transition to chaos, as well as the Cellular Automatons Theory and Neural Computers generated many novel and convenient mathematical models for describing the surrounding reality. Therefore, the Science of Complexity can be considered a collection of models and theories capable of allowing the understanding of local-global, part-whole type of relationships in a sufficiently general way so it can be applied to the study of all that is living, starting with genes, organisms and ecosystems and going as far as the study of transitions from atoms to materials and products, from computers to local networks and Internet, from citizen to group and society.

The Science of Complexity can be viewed as an integrating science, capable of ensuring an inter and trans-disciplinary approach (Nicolescu, 2008), to generate connections between different areas of knowledge, to create bridges between specialists in different research areas, from different schools and from different cultures, bringing them together in interdisciplinary teams targeting strategically important topics, such as those demanded by ensuring a sustainable development. The Science of Complexity is considered today the central pillar with which one can restructure information in a new and coherent paradigm that is comprehensible to all social levels, thus also having a catalytic role by accelerating the flow of information and knowledge to society. Furthermore, it generates the primary activities imposed by the joint Man-Environment evolution, a fact that traditionally has not been generally known or 'advertised' to the public at large since Evolution had been usually linked only with ecosystems and biological species, or – as a special case of the latter – to the appearance and gradual enrichment of the human intelligence along the ages. Nevertheless, one of the key contributions of the Science of Complexity towards a new paradigm is highlighting this entwined two-fold interdependent concept of Man-Environment evolution that is both reflected and provided by a constantly more complex and far-reaching design, control and utilization of products and systems, themselves with a previously unseen level of complexity.

Some of the industrially most relevant applications of research concerning Complexity today are:

- Miniaturizing antennas with fractal geometry (fractal antennas) for mobile phones- Motorola was a main leader in the research of such applications;
- Early cancer diagnosis using fractal analysis of a tumour made by a neural computer- the first patent in this field was obtained by the Health Discovery Corporation;
- The non-traumatic resuscitation after a heart attack through technology provided by "chaos control"- Philips carried out research for such an application;
- The super secured transmission of information through the synchronization of chaotic oscillators (chaos communication)- The military and telecom companies are at the forefront of research in this direction;
- The evaluation and improvement of performance in organizations through the application of techniques and simulation methods specific to the Complexity paradigm- counselling in Human Resources of some prestigious companies such as PriceWatherhouseCooper.

The few applications listed above, although chosen randomly, have all the same common denominators. Thus, they exploit a few key aspects which are traditionally either studied separately, or incompletely or (sometimes) are neglected altogether. These essential aspects studied (among others) by the Science of Complexity are:

- The non-linear properties of matter (or of the interaction matter-energy),
- The extreme sensitivity to initial conditions,
- The auto-similarity of spatial and/or temporal structures,
- Revealing new facets and meaning of the resonance concept, e.g. the chaotic resonance (which deals with the synchronization of chaotic oscillators), and
- Studying processes of formation/emergence of systems which exhibit auto-generated criticality.

The collection of models, theories and measurement and control techniques provided by the Science of Complexity enables a new way of perceiving and understanding Reality, seen as a structured network of complex and interdependent systems that evolve far from thermodynamic equilibrium may manifest coherence on a large scale, and for whose study a special, holistic approach is required. Such complex and interdependent systems coevolved, displaying phase transitions, spontaneous restructuring, and even generating new structures that are neither the sum of, nor the exact image, of their 'parents'. Consequently, it has been notoriously difficult to study, let alone to predict, the behaviour of such systems by using classic, reductionist methodologies. Additionally, and most importantly, the Complexity paradigm can be considered an important step in understanding Life and the relationship between Living & Non-Living entities/systems. This is achieved by understanding the structuring role of recursive processes that can model feedback loops and implicitly, auto-regulation processes. The negentropy production, generated by living beings through metabolic processes that ensues the homeostasis of living systems, is put into balance with the implacable trend towards entropy growth, the fundamental property of non-living matter. This “competition” between order vs. disorder, information production vs. entropy growth, etc., induces a special dynamic in Living systems that develop on a non-living 'substrate', and this non-trivial dynamic must be studied using a special methodology. The Living entity-Substrate network/assembly forms a new system that cannot be understood only from a thermodynamic perspective, using only very complex models based on systems of differential equations to simulate the behaviour of such dynamic systems. Nevertheless, traditional methods still play a part in the approach used by the Science of Complexity, but they are supplemented by an array of other methodologies that are used simultaneously with the traditional ones. These novel techniques strive to understand the streaming of information that ensures the regulation, adaptation, synchronization, and commencing of metabolic or behaviourist “procedures”, as well as the energy and matter flows that ensure the “functionality” of Living systems.

While the traditional approach that studies the energetic-material aspects/interaction(s) is a quantitative one, the study of information streaming (codes and messages) is a qualitative one and it defines a class of interactions specific to the Living entities/organisms. The intensity of energy fluctuations is not paramount in this informational connection, but the structure of the fluctuation is, meaning the message. From this perspective, the identification of the differences between a normal energetic-material fluctuation (noise) and one carrying information (message), requires a special approach, capable to identify codes (discrete elements) in a continuous fluctuation, the way of creating messages (syntax) and, naturally, the meaning of all of these. In this way, the study of continuous/discrete transitions, the identification of patterns (semiotics) and the decoding of meanings (semantics) become very important and are the centre of attention in the Science of Complexity. As a conclusion, for describing and understanding the evolution of a complex system generated by the continuous exchange of matter, energy and information between Living beings, which are also structured in their own specific way in atrophic chains that make in the end the GAIA network (Maiorescu, 2001), it is necessary to use from the classical methodology only the part concerning general issues of thermodynamics and allometry and at the same time set-up and structure a new science based on Information. Such a novel 'Infordynamics' science allows a major change of the way in which we perceive technology today, bringing it closer to the “production” means of living Nature (primary clean technology), and generating artefacts able to ensure the geostasis of the GAIA system.

2. From Quantum Physics to Orthotechnology

“Not everything countable is significant
and not everything significant is countable”
— Albert Einstein

The study of the consciousness (Chalmers, 1996) and the cognitive processes has been invigorated by the advances in computational science, structured around the central concept of *information*. Despite the daily overuse of the word *information*, one can state that its real meaning is not fully stabilized. Its intended meaning can be deduced only from the context in which it was used, from "information" regarded as (trivial) news, to a structured semantic meaning: "information," that is, to give a shape, to organize, to reveal a structuring pattern. The roots of the central interest for the deep significance of the "information" concept are to be found in *its relation with artificial intelligence, with cognitive processes, with the analysis of* mind-matter *interaction.*

One can define:

- *Information in Shannon-sense* is what eliminates or reduces a given, *a priori*, uncertainty. It does not bring an absolute additional knowledge, except for ameliorating an initial state of "accepted ignorance." The concept is defined from an anthropomorphic observer perspective, and as such certain subjectivity is implicitly involved. The emphasis is set rather on a mathematical framework allowing a quantitative approach originating from an existent, well-defined, uncertainty.
- *Information in Săhleanu-sense* (Săhleanu, 1973) originates from the remark that in nature some interactions cannot be formalized through a classic causal interconnection, based solely on energy and substance fluxes, together with a conservation law. Săhleanu defines an *informational type link (ITL)* influencing the dynamics of a system. This happens not through intensive variables, but rather through special fluctuations acting as "messages;" they generate new aspects/mechanisms like storage, decoding, tuning or triggering. The ITL aspect is revealed in the context of the generalized hierarchical systems theory, where the global stability of a system is accomplished through formal and informal cooperating mechanisms of the constituent subsystems. Each subsystem has a certain structure, organization, and a sensitivity to information defined as *informability*, that is, its property to react to specific patterns of energy fluctuation.

It is clear from the previous remarks that the concept of information is strongly connected to structure and organization. From this perspective, the suitable methodology is rather close to grammar, semiotics and semantics research fields than to the study of dynamical systems - mathematically defined as sets of differential equations (Hoffmeyer, 2008). *While the energy is a measure of the dynamics, the activity strength of matter, the information is a measure of its structuring.*

The informational connection is more subtle and difficult to reveal, as it appears through correlation factors at different scales; the stronger the irregular or the unstable character of the system, the more pertinent is the manifestation of the informational link. The main classical methods for the theoretical and experimental analysis of fluctuations and irregularities rely on statistics, and are unable to decipher and characterize the structure and the related informational link. It is therefore necessary to initiate a fresh approach in theory and experiment, based on a new ontological reality model. *Ortophysics* (Drăgănescu, 1985) is exactly such a model. Defined in 1973 by Mihai Drăgănescu, the orthophysics model of reality relies on combined structural and phenomenological vision of the world and leads to the definition of so-called *The Ring of the Material World (RMW)* (Drăgănescu, 1979). It postulates the existence of a deeper layer of existence, named *orthoexistence*, acting as source and substrate for any physical universe characterized through energetic and material (substance) properties. What exists is no more than orthoexistence and universe, in a subtle connection linking entities with distinct fundamental properties. Orthoexistence, as substrate of the physical universe, is made out of two components, *lumatia* and *informatter*. Lumatia represents the substrate of the substance, without any self-structuring capabilities, while informatter is the component responsible for structuring lumatia. In the absence of informatter action, lumatia presents a trend toward global equilibrium and stationarity; it can be regarded as the source of energy in the (physical) universe. The structuring characteristics of the informatter reflect in the concept of phenomenological information that manifests as physical "sensitivity." These processes within informatter, named *orthosenses*, induce a special dynamics in the orthoexistence. When the two primordial ingredients (lumatia and informatter) couple, the matter is structured and maintains this state through orthosenses; the result is the generation of universes with specific physical laws. Part of the informatter remains unstructured and potentially available for new structuring processes. The informatter is not conscious and does not contain intelligence, but is nevertheless an essential ingredient of any living organism.

Living organisms have (direct) access to informatter through a specific interaction named *intro-openness*. This interaction endows the living entity with distinct phenomenological attributes: sense, psychological continuity and the quantification of the spatiotemporal reality. Orthophysics is the science having as its goal the study of the orthoexistence and its properties. It tries to elaborate a model of the depths of the material world that explains the way orthoexistence is coupled with the universe, from the formation of the elementary particles to a unitary description of (quantum and macroscopic) physical laws, biological phenomena and the mental and psychological processes.

The structural and phenomenological perspective of the world proposed by orthophysics goes beyond the limitations of the present contemporary science approaches, placing in its centre a common principle for all the phenomena studied in the natural sciences and information theory.

Informatter (the deep "sensitive" substance) *has* orthosenses, *generates* orthosenses and *searches* for new orthosenses. The material world evolves in time and space under the "pressure" of a fundamental orthosense: "*to be*". It manifests *within* (giving the existence a stable unity with its profound source), *outward* (generating fundamental secondary orthosenses) and *inward* (the orthosense of searching of orthosenses – the evolution of the physical universe and living nature toward accumulating new senses, perceptions).

The phenomenological information cannot be formalized, but has a structuring action through the relations it establishes between senses. The structuring of the orthoexistence appears when the structuring informatter couples with lumatia; this results in a stable non-living universe fully describable through mathematical formalisms. Its coupling with lumatia is the source of its stability. On the other side, the living universe presents a phenomenological intro-openness, which is not possible to describe through mathematical formalisms.

Life (the animate matter), is generated from the same ingredients as the inanimate matter, but possesses supplementary features on the information side. In the non-structured informatter there is a category of orthosenses able to attach only to structures with a level of organization beyond a certain minimum threshold (resulting from the interactions between the elementary particles, through electromagnetic forces). These orthosenses are called *orthobionts*, informational components setting a characteristic, *specific unit* to structures of minimum complexity. As a consequence, elementary forms of life begin with an "intro-open" molecule with reproduction capabilities. The evolution towards the next level, living cellules, is initialized through a process of association between these primitive "animate' molecules. The living matter, through its intro-openness, subordinates and controls the properties of the non-living matter, in a trend to evolve and towards increased complexity.

Every living entity has mental processes. In the elementary living matter these mental processes flow without intelligence or consciousness. These two last attributes emerge only at the level of organisms possessing a nervous system.

Intro-openness acts only through animate (living) matter; the mental and psychological level of the physical world is the necessary interface between the spatiotemporal universe and the profound substrate, and allows the birth of ideas. Ideas cannot exist in the profound level, for their substrate is the nervous system in the physical world.

The *structural information*, the only one analyzed by science so far, is in fact of secondary importance in the physical world, relative to *phenomenological information*. Orthosenses from informatter are the basis for the generation of physical universes, but they are phenomenological informatter.

The most general form of information has a dual formal-phenomenological character. The formal aspect contains structures specific to syntax and semantics (either reference or in the specific context), while the phenomenological aspect contains the associated senses or orthosenses. The mental and psychological information of the human consciousness is presented in this dual form. Obviously, there are cases where information appears only in a formal structural and syntactic way, as approached in the classic theory of information transmission or in bioinformatics (genetic information).

Consciousness operates with structural-phenomenological information, thus escaping from any mathematical formalism approach. The structure of the brain allows a bidirectional communication between the structural and phenomenological aspects. RMW model postulates three levels of interaction: phenomenological interaction (to be), structural-phenomenological interaction (to know) and the pure structural (to say). The nature of man places him at the border between existence and orthoexistence. This enables the dual capacity to interact with the informatter (from the profound reality) through intro-openness and with the physical reality through openness (senses). This dual interaction can be developed in a conscious manner, through mental and psychological process, leading to a control of the reality. Man is able to "interrogate" and "control" domains of physical reality through direct physical interactions (forces, energies, moments). A second alternative, at a subtle level, is to act through inter-openness upon the mechanisms in orthoexistence responsible for the generation of that specific sector of physical reality. In other words, man and society are able, through knowledge, to actively influence processes in orthoexistence. The results can be materialized in the actual spatiotemporal universe, generating anti-entropic effects or to even lead to the generation of a new universe. A new technology, called *orthotechnology*, will assume the development and use of a new category of equipment: *orthotrons*. They are intro-open systems without consciousness, analogous to present instruments and tools from the lower levels of the physical reality. The intro-openness could thus provide a justification for the effects of the interactions between mind and physical reality. These effects have been observed so far accidentally or in empirical experiments trying to quantify the influence of human intentions on the behaviour or properties of a designated physical system.

3. Mankind's Spiritual Becoming and the Birth of the Awareness Society

"The true meaning of History in a Universe is to induce/aid the Becoming of Consciousness"
— M. Draganescu

The orthophysical model of Reality also postulates the existence of a *Fundamental Consciousness* (FC) whose appearance and operation takes place via the primordial ingredients of the 1st level of Existence. Therefore, FC itself becomes a primordial element, situated in non-space and non-time (i.e., in orthoexistence). The individual and group relationships with FC define a special type of behaviour for the Human Being(s), namely the *spiritual behaviour* (Drăgănescu, 2003).

The scientific approach can recognize, within the framework of a transdisciplinary ontological vision, the existence of the FC; however, it would not be capable to principally experiment it directly by using its specific methods.

The *Initiation* of the Human Being is a special process that would allow him/her to access this spiritual dimension, and which has been repeatedly outlined along the ages in many esoteric or religious cults and practices (mysteries). Since the relationship between the Human Being and FC is a phenomenological one, being mediated by states of *being* that – by definition – cannot be expressed fully or exactly in a formal language/manner, automatically implies that an authentic initiation does not require or consist of a rational accumulation of concepts and theories, religious or of any other character, but by a personal experience and a deeply felt communion with the inexpressible.

According to the orthophysical model that postulates the structural-phenomenologic nature of the Universe, one can identify three types of interactions in orthoexistence:

1. Phenomenological interactions (states and processes of 'being'),
2. Structural-phenomenological interactions and operations at the level of "knowing", and
3. Structural interactions formalized at the level of "saying".

A model of the interaction between the individual consciousness and FC based on RMW states that *the key point of a Human Being's Spiritual life is the emotional trace, or 'imprint', left by his [state(s) of] Being in his Affect*[1]. The Human Being thus connects to the FC who will transmit her/him "tasks", in consonance with the overall harmony of Universe's evolution, and this connection leads to the awakening (via the Affect) of internal energies and desires of search, creation, fulfilment, and embodiment in the Material of the elementary cell, or 'brick', which (s)he, the Individual, can contribute with at the construction of Reality, according to her/his own particular psychophysical features. This state "feeds" the intellect, which will *act to adapt the "intentionality seed" (brought in by the [state(s) of] Being) to the spatial-temporal requirements of the physical Universe.* The intellect, as semantic processor, operates in the Structural and generates algorithms, technological processes, inventing materialization solutions. In other words, a subtle interaction (mediated by the Affect) is forged between the FC and the individual Human Being who is 'immersed' in a spatial-temporal reality.

The mind of an individual, when in a neutral state (i.e. with an attitude without any intentionality, in a contemplative state), can be configured such that it could "receive" its own personal Being process. However, the message is typically received in an extremely hermetic language, structured according to the laws of a very different medium/environment, which knows no spatial or temporal boundaries or 'milestone' delimitations/demarcations. Therefore, decoding the symbols within, and the meanings of, this message, as well as its adaptation/conversion into the physical reality by material implementation assumes a perfect synchronization and cooperation between the individual's affect and intellect, a very good control of the affective/emotional states and a developed capacity to formalize and operate logically with abstract notions specific to the tangible surrounding physical reality.

The abilities required to operate coherently at the interface between "knowing" and "saying" are formed by an academic education, which develops the rational, logic, intelligence originating from the brain's left hemisphere. This intelligence can be structured beyond the perceivable reality by further developing the self-awareness and filtering the existing knowledge by using the most refined paradigm(s) currently accepted

[1] Merriam-Webster's on-line dictionary definition of 'affect' is: 1- (obsolete): feeling, affection; 2- the conscious subjective aspect of an emotion considered apart from bodily changes; *also*: a set of observable manifestations of a subjectively experienced emotion <patients…showed perfectly normal reactions and *affects* — Oliver Sacks >.

academically. The development and performance of this rational intelligence actually strongly determines the quality of the decoding process applied onto the "message" received via the state(s) of Being.

The abilities required to operate coherently and consciously at the interface between "being" and "knowing" are formed by gradual and systematic long-term application of various meditation, prayer, or other introspection techniques targeting self-knowing and exploring the individual's own affective dimension, i.e. elements that define the so-called "spiritual life".

The lack of such a spiritual life/development and the exaggerated role and importance attributed solely to the rational intelligence would transpose ideas in tangible facts and in material objects without any moral discrimination or other judgment based on a holistic vision of the Individual's true role/calling and/or of Mankind's/the Individual's necessary Evolution. This has led to the present-day society that clearly values mostly the Matter above the Human Being, i.e. the artefacts and the profit(s). The present-day paradigm for mankind's *modus operandi* can be summarized (even if somewhat simplistically) as follows: "An individual is valued on what he owns materially, and on either how much more he can further gain or his capability to be used to make others to gain". The one that has the highest number of possessions at the end of life 'wins'. Any other 'spiritual' elements, i.e. non-provable tangibly in a reproducible manner and non-formalizable, are -at best- either allocated a 'lateral' place as a 'distraction' or (e.g. religion) being partly necessary for mental comfort or a 'good' life, or –at worst– completely denied as possible or existent. However, it is clear that the existence of such a materialistic society that encourages consumption as main engine of 'development', that always seeks quantitative clear-cut estimators of success based solely on profitability, that demands constant economical growth as a primordial requirement of well-being, will clearly not survive in a very long-term (i.e. on a geological scale) due to either unavoidable exhaustion of all available resources, or overpopulation, or other causes (e.g. pollution or climatic changes). Moreover, this process will literally enslave the masses for the benefit of the few, will alienate most of the members of the society and de-humanize their activities, while at the same time also causing severe ecological unbalances.

We can conclude that modern Science, although has initially started along the route of explaining Nature's secrets (and which will still remains a key drive in the future), it now reached a critical point in which it can formulate a new vision about the entire observable Universe. This vision should be based on quantum physics, on the study of Consciousness and of the human Mind, and it could also define the role of the Human Being in a similar manner (although probably using very different words) with the teachings of the great Initiates of the mankind. Thus, through Knowledge and Affect, Man can attain Wisdom and awaken spiritually in order to subsequently assume a new role of "interface" between the observable/intelligible universe and the Sensible Universe, between potentiality and history. We shall exemplify with a quote from an Orthodox Christian sacred text that defines the role of the Human Being as central element of the Divine creation (i.e. of the Fundamental Consciousness). The saint Maxim The Confessor said:

> The Lord lastly introduced the Man among the other beings, yet above all beings, as a laboratory that brings together the extremities of the universe through his own parts and thus brings into himself in unity all those that by nature previously had been split asunder and separated by great distances.

The Human Being has to

> unify the entire sensible creation *through knowledge and contemplation of meanings in things and love,* and then to unify the sensible ones with the intelligible ones (Larche, 2001).

The Man, himself/herself a part as well as a whole, an *arhem* and not a system, becomes the intersection point, the crux, between the physical and the metaphysical worlds.

> The Man is simultaneously both the conclusion and the apex of the creation. This is why he was created in the seventh day. As soon as he appeared, everything was complete and flawless, both the upper and the lower worlds, because everything is contained within the Man he encompasses all forms – says the Zohar.

For Pico della Mirandola

> The man is the link between the entire nature and an essence comprising all its saps, and this is why the one who knows thyself knows everything.

Similarly, Rudolf Steiner also asserted that

> The Man is an organism that transforms different forms of the force in the force of his thought, an organism which we maintain active with, what we call food, and by the means of which we can produce, what we call

> thoughts. What a marvellous chemical process must be that that could transform a simple quantity of food into the divine tragedy of a Hamlet,

and he concludes by saying “The Man is placed in the middle, between God and the animal”.

Although mostly in an archaic language, all the texts above clearly highlight, from various perspectives and from different epochs, one and the same thing: The role of the Human Being as a thinking being, as the ‘alembic’ or ‘retort’ in which is synthesized the conscious reflection of the Universe and thus making possible, beyond/above a certain developmental level, to perceive the FC. It must be outlined that the above statements have been mostly the results of an intuitive, revelatory process, based on successive refinements of expressing special states of being (altered states of consciousness triggered by prayers, meditations, and shamanic practices) in a formal and verbal language.

We can thus conclude by saying that the Human Being, as a special self-aware being capable of interacting creatively in and onto the physical universe by the means of artefacts, is an Arhem equipped with a structural-phenomenological “processor”. A key feature of the Human Being is thus his/her capability to interact with the phenomenological by means of the intro-openness; another one is the capability to participate personally and directly in stabilizing the Universe in which (s)he is immersed. Through knowledge and assuming the responsibility of its usage, as well as through individual and collective development and evolution *the Human being becomes the keystone for the quantum stability of this Universe*, and in this role (s)he fulfils her/his destiny.

This development level that assumes the spiritualization of the entire human race is accompanied by the emergence of a new specific social structure: The Awareness Society. This will be a spiritual society, not in a passive, meditative meaning, but rather denoting a society in which the education of both rational intelligence and the spiritual evolution of the Human Being become well-defined targets. This would be an informational, scientifically, technologically and agriculturally active society, seeking to understand and assist the Evolutionary process of Life on Earth, and in the Universe as a whole. Only this type of society could possibly ensure a sustainable long-term survival and evolution of the human race.

4. Conclusions

"To be so bold as to give people the conscience of greatness that they themselves cannot imagine."
—*Malraux*

The exponentially growing dynamic of the modern socio-economical life, in the context of globalization and climatic changes, puts an enormous amount of pressure on each individual. Trying to survive in this turbulent and increasingly stressful environment in which the prediction of future events is becoming more and more difficult, and requires an increased effort to ensure both continuous adaptability and, at the same time, also to provide the ability to deal efficiently with a possible failure.

It is common knowledge that every critical point (either a dilemma or a choice) requires a (sometimes hard) decision that subsequently results, or leads to, a change. The psychological tension induced by the uncertainties that naturally accompany any decision in a primary unpredictable evolution has the power to affect human beings at both an emotional and mental level. Presently, we, i.e. humankind, are lacking an appropriate 'bird-eye' vision of this dynamic evolution of Reality, thus missing not only the entire picture but being incapable of understanding its dynamics, its sense/direction of evolution, or its specific manifestations in various sub-systems. This is largely due to the fact that we are still using concepts and information that had been specific for a previous, much more static, stage which admitted a *linear estimation* and a *Newtonian approach.* However, nowadays modern (wo)men deal harder and harder with stress at increasingly more numerous levels, a fact illustrated by the continuous world-wide degradation of the quality of life and by the negative decrease of the immune system's performances in general as well, with dramatic consequences in the continuous degradation of the physical and psychological state.

A solution for improving the individual state of health and the growth of the chances of survival of the human species in the context of the actual socio-economical turbulences, would be accepting on a social scale an immediate need to begin taking the following measures:

- Understand the processes which mankind experiences today (a "forced" transition from the *industrial society* to the *informational* one, that of *Knowledge* and *Conscience*),
- Spread the study, understanding and application of the Science of Complexity. Only propagating and using this new paradigm into the society would enable us to identify and foresee certain patterns in the

socio-economical turbulence, thus providing a step upwards/forward to a conceptual stage superior to that of the current paradigm,

- Refine and adapt formally, at the understanding level of the modern society, of the clerical information and advice, which assume the placement of the Human Being, through faith, education and triggered self-awareness, in the centre of the Universe's evolution (eco-sophia),
- Develop a new educational system based on the unity of knowledge (interdisciplinary vision) and the understanding of the cognition processes (the science of the mind) in harmony with the assumed role of the Human Being to be, first and foremost, an unique spiritualized entity capable of creating adequate conditions for the protection, maintenance and further development of Life on Earth.

INNOVATIONAL CULTURE THROUGH PERFORMANCE AND INVENTICS SCHOOL FROM IASI[1]

BORIS PLAHTEANU AND MIRCEA FRUNZA

1. Premises, Commodities

It is questioned less whether the historical existence of the human being relies directly on the ability to turn knowledge into creation.

Such a point of view allowed Pierre Janet to state that social life is subject to two fundamental laws: the law of invention and of preservation on one hand and the law of creation and repetition on the other (Janet, 2004, p. 83).

Submitting to these laws in the process of adaptation, cohabitation and survival, man began to understand that he has a potential too great to waste. "To have ideas and waste them is indecent" notes Mircea Malita (Malita, 1999, p.13-19).

As such, to avoid similar "indecencies", the education and stimulation of the creative resources of all the members of the human community is a prime objective.

In a competitive system the value of human actions evolves continuously as manufacturers are restless to bring new products on the market.

They can be outdated by better projects, new materials and effective technologies. A good performance of the product is assured by programmed action, innovation based on patents, know-how and protected by products of industrial estate.

It is necessary to continuously research costliness, by rethinking objects, reassessing materials, introducing new technologies and creating new products that match some innovative requests.

---

[1] National Institute of Inventics, Jassy, Romania - www.inventica.org.ro

In this context there is the requirement of a new instrument based on methodology, and at the same time technical and economical as well, which will be able to guide all the processes to the optimal solution. More than that, it is necessary to define a strategy and a new line of thought for the analysts to follow up throughout the entire life-cycle of the product.

These purposes are achieved in an interdisciplinary way, in which the various scientific approaches are governed by a philosophical perspective that ensures the conceptual level and the unity of the discourse. The philosophical approach itself is practised in several specific fields and at various levels, being used in order to lay the foundations of a new domain of research. Specific theoretical problems are elucidated through concrete examples furnished by sciences and discussed from the perspective of inventics. Graphical representations help in giving a systemic and integrative perception of complex issues.

Philosophically, the research program is relevant in that the analysis of technical artefacts poses interesting new challenges for existing approaches in various philosophical disciplines; for example, in epistemology (functional knowledge), in philosophy of science (theories of function and explanation), in the theory of action (the agentive function of artefacts), and in ontology (the constitution of artefacts).

Creative designing along with costliness engineering fulfils this objective in a systematic and conceptual way of action, maximizing the ration between the functional value of a product and its cost.

In this action, the "oxygen" of creative designing is functional analysis and application of the methods of inventics.

Costliness engineering also contributes essentially to this intercession by being a method of organizing engineering and economical activity which is based on techniques and procedures that ensure a rise in creative performance and effectiveness of the operators in this activity.

The objectives and problems of creative engineering may be:

- Raising the performances and quality of products and their ability to compete,
- Reducing fabrication costs,
- Reducing the amount of material, energy and workload,
- Predicting accurately the development of technology,
- Rising production rate without significant and important investments,
- Finding solutions for technological strangulations,
- Replacement of unusual materials and supplement subassemblies,
- Detecting, reducing and removing wastrels,
- Improving the ecological indicators of production.

According to theory, any object has reserves to improve his resources. There may be:

- dissonance between the object specifications and his functional and necessary indicators in the real working conditions.
- ineffective use of science and technical information
- insufficient information of the operators on new constructive solutions, on utility of materials, on prices and deficits.
- technical conservationism of the specialists
- weak cooperation between specialists who are specialized in narrow fields.

Even by triggering the perfecting mechanisms, after some time, as a result of new technical and scientific developments (discoveries, inventions, new materials, technologies, administrative methods) the product becomes obsolete and more improvement actions becomes necessary.

Innovation becomes vital if confronted with fierce competition that cannot be surpassed other than by creating super products; an objective impossible to reach without an increase in the amount of activity in technical creation. “Innovate to survive”, is a concept launched in 1980 by the president of the French National Council of Patronage: “Those who miss technological corners are bound to vanish or to become peripheral” (Chenevier, 1980, p.42-45).

Under these circumstances, innovation is no longer an academic debate or luxury, but a battle for survival.

The drawing in figure 1 shows a representation of the industrial-economical curve of Kondratiev, drawn in 1925, that shows the periodicity of the economic crises, that occur once every 50 years, and *the innovation curve* drawn by the Japanese researcher Tosio, for emergence from crisis.

Indeed, Tosio found that where the Kondratiev line shows minimums, representing economical depressions, the innovation curve shows maximums, a fact that is explainable as follows: under the pressure of a crisis man uses all his creative possibilities to overcome it (Belous, 2005, p.13).

But intensifying the creative processes is determined by a second group of convincing arguments: the rapid growth of the complexity of technical systems and, implicitly, the number of the problems that need to be solved by creative constructors.

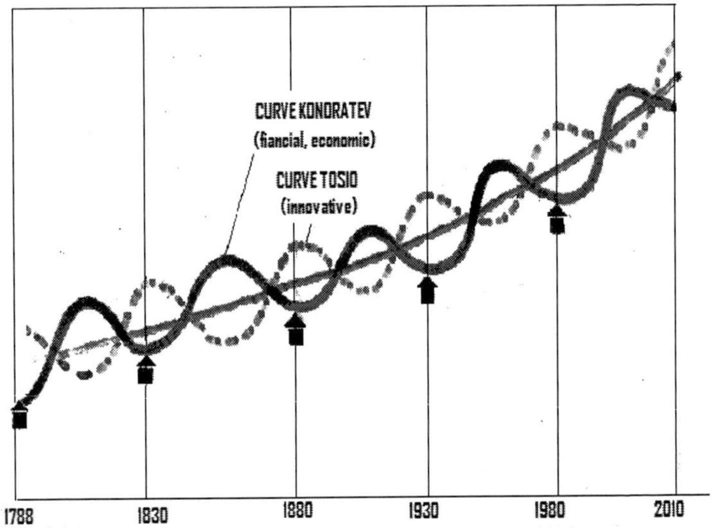

Fig.1.

An invention brings along progress, but more important to the society is the education of the inventor.

An invention is only a product, while an inventor is the means to produce it, capable of increasing the number of products and their quality.

A complete engineer is characterized by four fundamental attributes:

1. knows the technique
2. applies the technique
3. operates with technique
4. creates technique

The traditional academic technical education (still practiced in many places today), focuses on forming an engineer that meets the first three characteristics, the creative side being played down as an activity that should happen spontaneously. The structure of the academic courses often creates inertia and psychological jams to creative thinking.

For Romania the integration in the European Union cannot be done only by raising the human activities to those of the advanced European structures, but from the specific Romanian contributions, in areas that they have an advanced place in the world.

One of these areas is *inventics*.

For the training of an inventor in the first way, is necessary as a new scientific discipline. The name *Iasi School of Inventics* was given by the well known French specialist in combinatorics, engineer and mathematician Arnold Kaufman, "*L'inventique*" (Belous. 2005, p.16).

The Romanian concept of *inventics* is polyvalent and it includes science and art in an optimal way, their creative synthesis, logistics and algorithms of the heuristic ways, established as praxeology:

- finding of the domain of the creation
- finding the problems of the actual level of the techniques, defining of the product performances.
- discovering the solution which assures those performances, till the industrial implementation of the original solution.

The development of such a science, i.e. *inventics*, began in Romania in 1967, when in Iasi a team of scientists and academic professors organized the first post academic courses of technical creation.

An independent system emerged of the processes of invention, forming as a collective lead by the famous researcher and professor, Vitalie Belousov (Plahteanu, 1999, p. 46).

2. The Invention School of Iasi

The Invention School is:

- based on the analytical study of the factors in creativity;
- learning from the positive results obtained through methods of stimulation of the creative capacity;
- based on the experimental data gathered over more than 30 years;
- and has demonstrated that it is possible to develop a pedagogy of creativity on every step of education and that no matter how mysterious the connections between the continues and the sub continues are between one's intellectual and emotional sides, based on which the personality of the inventor is structured, the basic personality traits that encourage invention can be developed by education, by an engineering of technical creativity.

3. Study Programs and Strategies

There are two ways to increase the creative training of the students: either through introducing a special creativity object, or through complete conversion of the way to teach almost all lectures.

The basic role in preparing the engineers for creation must lie in the academic teaching both by converting the specific lessons in a creative way and by introducing courses of technical creation or creative design.

This goal was achieved by Gheorghe Asachi Technical University in 1981 and has spread to other universities in the country.

In order to increase IP importance and to improve creativity culture inside the university, a human resource development program proposed an objective which could meet these requirements.

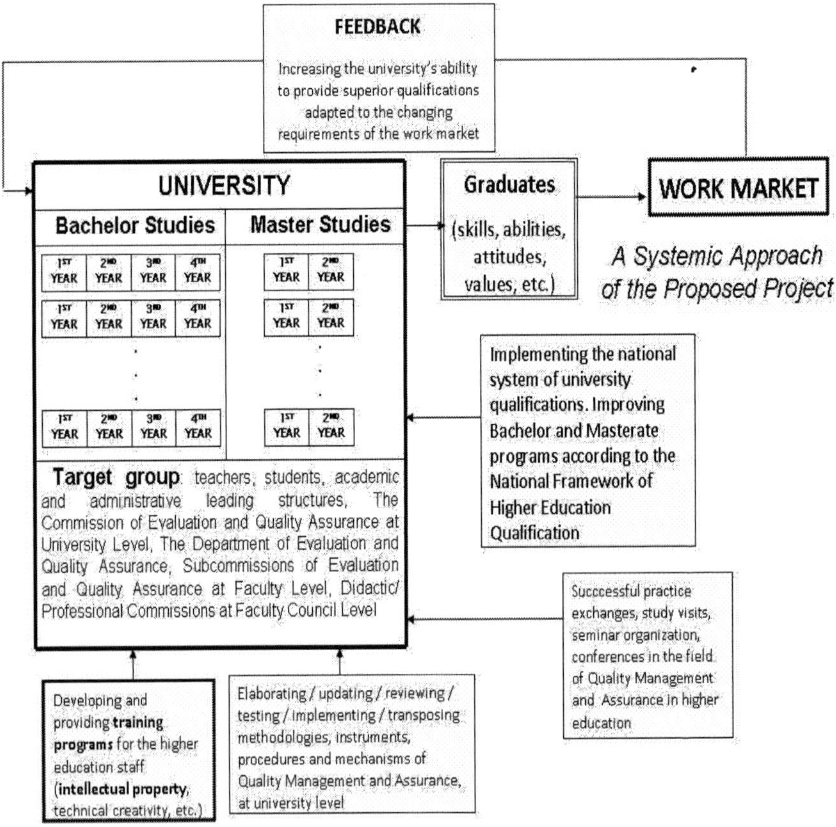

Fig. 2. Industrial Property Promotion in the strategic plans of the university and future objectives

Over the years there were many post academic courses in invention that were completed by over 2100 engineers and high school teachers in which theoretical and practical knowledge was spread.

The actions and achievements of the invention school represented the support for the birth in 1992 of the *National Inventics Institute* of Iasi.

Developing the strategy of research in invention, developing the logical foundations of creation using algorithms in the heuristics, developing procedures, techniques and logical methods of creation, the Institute is involved in an intense activity to train new inventors.

In over 19 years of activity, the *National Inventics Institute* of Iasi promoted, on countless occasions, actions to support the presentation of one's own scientific results and offered adequate space for the cooperation of specialists in invention, technical creativity and innovation.

As a result of these initiatives new strategies for research were developed, and collaborations and programs were launched in new fields of work.

In 1999 the Technical University Gheorghe Assachi, Faculty of Machinery Construction started the masters course *Creativity and Technical Performance.*

4. Educational Strategies in Innovation. Innovational Culture and the National Inventics Institute

Using the entire construction developed by us in inventics, and the systematic approach to the complexity of forming such an innovational environment, allows us to point out that the solution to this problem is a string of projects of content and management:

- Determination of the place of innovative culture in general culture, and its influence on crucial areas of human activity (economy, law, leadership, education, politics)
- Creation of solutions to previous problems, the informational and instructive system, which will be applied in the educational system.
- Developing a set of recommendations to governmental institutions, popular structures with the purpose of promoting innovation culture
- Raising the level of innovation culture as a factor of evolution.

The process of innovation development has two main components:

- realization of innovation projects;
- development of the innovative potential.

In our opinion the essence of this stage is illustrated in the concept of "innovation development". The justice of approach requests precision, in the assembly which assesses the essence and the mechanisms of the innovation development.

The novelty itself, even scientific, by the technical - scientific development, became an innovative product As such, an innovation cycle begins in the first step of scientific research, of conception, design and construction. The results obtained create that fertile space in which the innovation activity begins both in real economy and in other domains. Innovation development has to bear a complex character as we understand it primordially as a chain of novelty. It has a greater success when it not only comprises of a small domain, but multiple areas that influence the final result (leadership, marketing, instruction, finances, etc).

The Relationship between Quality Culture, IP Culture and Creativity Culture is shown in Fig.3

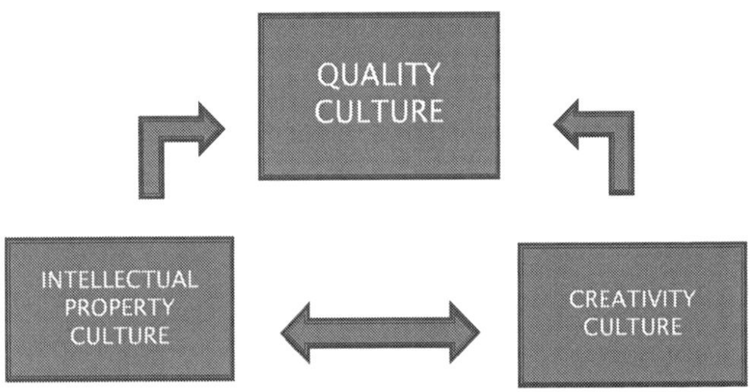

Fig. 3.

The idea can be potentially innovative when there is the certainty that, by crossing the scientific - technical stage, and through the innovative cycles, an innovative product will materialize.

Any innovation development is not only an innovative process, but the creation of a system of factors and conditions, necessary for its completion – so an innovation potential.

The absence of any landmarks regarding the innovative-inventive instruction of the members of the companies and institutions of scientific research makes the innovative activity impossible.

To this we add the construction of a system, a problem that is crucial for the entire innovation activity: the promotion of the protection of industrial property.

This is the place from which the problem derives: the measurement of the entry parameters of the second component – the innovative potential and establishing its place in the general potential of the company, school or leadership department.

5. The Synergic Academic Space, Technical University – National Inventics institute – Regional Centre for Promoting the Industrial Property Iasi

All this has created the premises for the birth in 2001 in this area, by means of an agreement, to the National Office of Inventions Iasi which became a part the Romanian Patlib Network.

In this crucible of the University, The National Inventics Institute and the Regional Centre, young inventors, through their creative activity, became the catalyst of novelty in the university and then in the industrial space that they work in. The decision to introduce, in the teaching plans of the "Gheorghe Asachi" University, the course of "the basis of technical creation" lead to the substantial growth in the number of propositions for inventions registered at the National Office for Inventions and Trademarks. Year by year the invention school from Iasi realizes an important production of student inventors. The Technical University Gheorghe Asachi of Iasi realizes one of the greatest densities of recorded inventions in the country.

The participation of these institutions as a group concerns, in all phases, the transfer from the idea into industrial product in these consortia: University – National Inventics Institute – Regional Centre, assures the synergic space for growing the quality of the inventive act.

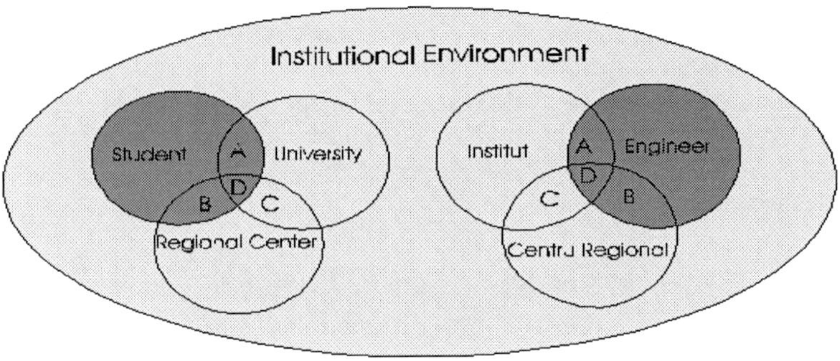

Fig. 4.

Zone A - the interaction between the individual and the university institutions or the Invention Institute matches the engineering domain of creativity.

Zone B - of interaction between the individual and the subsystem regional centre, thus contributing through creative potential and expert knowledge to industrial property.

Zone C - between the teaching institution and the regional centre means efficient organizing and the capacity to utilize effective development strategies.

Zone D - integrates the sum of the subsystems and can be considered a zone of maximum creative efficiency, a "transcreativity". An effective strategy in developing systems has to amplify this area in order to make the whole act of invention effective.

In this synergic system it is necessary to establish the concrete content of each factor and develop typical organizational patterns, to create an environment in which innovation potential can form, taking into consideration the characteristics of the region we work in.

This is about, for example, the solution to the structure and organization of the activity of analyzing and patenting

Given the important role of the innovation factor and the insufficient training of many specialists in the problems of leadership in the innovation world, the basis of the model is to be presented to companies as a real form of support from the national institute of invention.

Understanding exactly where, how, and in what succession the efforts should be directed, will allow, in the area of innovation, a union of the possibilities of the companies and those of the regional institutions. This way the possibility to eradicate routine and effectively use the new achievements, inventions and know-how will present itself. Without large costs we will succeed in mobilizing a huge resource, which now is neglected. The strategic resource of the new century is innovation culture.

World experience shows that surpassing innovation stagnation through investment only is not effective. According to the opinion of European experts, written in 1995 in the green book, the state of innovation in the European Union is unsatisfactory. This has a lot to do with companies avoiding risk and with multiple barriers to creativity: bureaucracy. This way we find that many problems come from other areas beside the financial one.

Still it is obvious that the training and educational resources alone won't be enough. To radically change the situation it is necessary that the state supports the politics and initiatives in innovation culture; there is an unlimited area for the innovation culture to manifest: from creating conditions for the creative potential in the interest of the society, to ensuring a balanced process to reform it.

And the results in this collective effort are obvious: there are comparable results seen in the two histograms of the evolution of the number of patent request by the students of the Technical University Gheorghe Assachi of Iasi and the density of patent requests per 100 thousand people in a comparative analysis, in Romania, in assembly Bucharest, in Iasi, and in University.

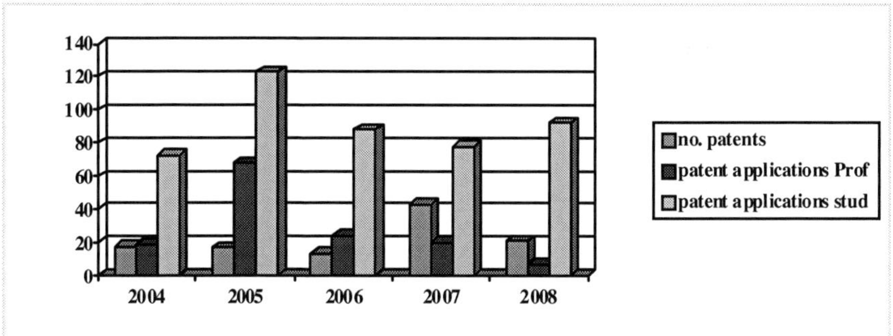

Fig. 5. Evolution of the number of patent request by the students

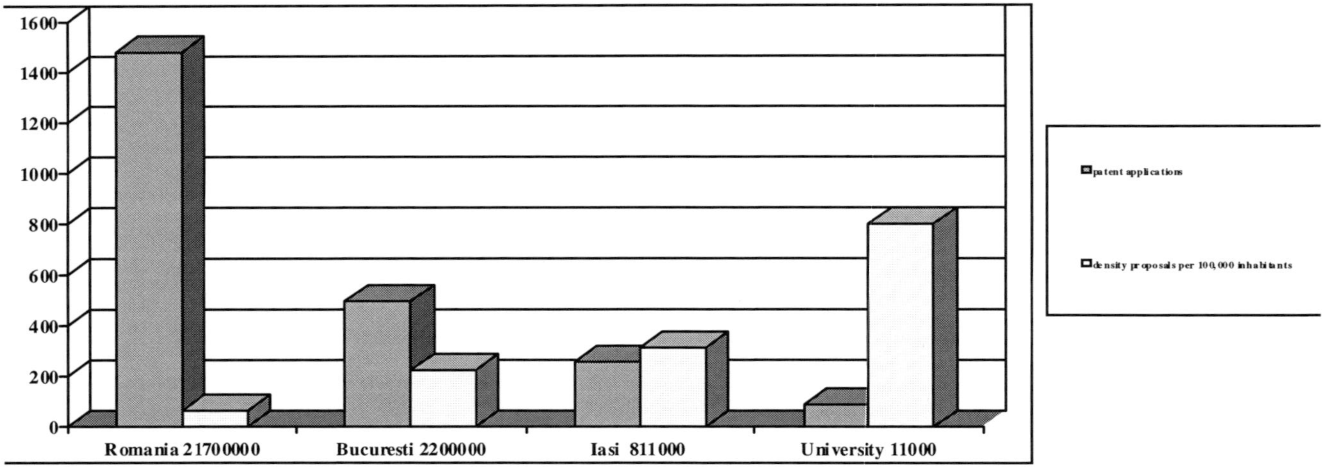

Fig. 6. Density patent applications per 100,000 inhabitants

We add to this approach the scientific manifestations organized by the Institute for capitalization of the research results: national and international forums, which highlight the level of performance and quality of the new products and technologies from all technical domains. These forums must involve a real knowledge exchange, experience and results regarding the conception, design and advanced construction of machinery, equipments, technologies and fabrication systems.

In this way a reflection space is realized, as well as evaluation and a good defining at the new concepts of the research, development and realization in the field of engineering in general, offering to the specialists the opportunity of profitable meetings and exchanges of ideas.

A short look over the deployment of creativity and performance shows that at the beginning of the new millennium we are witness and at the same time the actors of a new thinking at the global level. Through those central themes: no sphere – man – techniques.

The chances are postponed, as are the risks involved in this evolution, the inventive solutions and the challenge of the future.

Conclusions

1. Innovation culture reflects the general orientation of the individuals, based in knowledge, dexterity and adaptation, motivation and also in the way of conduct.

2. Training innovation culture is linked, above all, the development of creative ability and creative potential of individual achievement.

3. With a highly innovative culture of the society by force of the inter-correlations and interdependence of its parts, one of the components attracts a quick change to another.

4. In the conditions of a modest innovation rate, a strong management and legal impulse is necessary for self-regulatory mechanisms to operate, for development in an organized process, governed by a determined dependency structure, rules of conduct, responsibilities of the participants.

5. It sunken theoretical assumptions on innovative culture, his scientific components highlighting the favoured and restraining factors of innovation culture.

6. An important role here can be played by the world: regional international inventions Forums.

7. Setting the future is in our minds and hands, and in this world, smart ideas must be used as a resource that cannot be used up.

SOME PHILOSOPHICAL ASPECTS REGARDING TECHNOLOGY

VIRGIL MOLDOVAN

During the last century, men have been witness to the intensification of automation and mass production. Every day life was transformed, by transport and telecommunication achievements, from the car to satellite communications. Equally revolutionary was the development of some materials.

In contrast with the macro scale, with a lot of innovations of the 20th century, the technological frontier at the beginning of the 21st century is at a microscopic scale. The imminent discoveries from this domain and the different applications of nanotechnology facilitate the emergence of more rapid and "intelligent" computers, of some news realisations in medicine and of bionic implants in miniature.

The new science of *genetic engineering* involves the manipulation of one organism's genes and can improve the quality and the quantity of agricultural production, having applications regarding disease prevention and treatment.

The digital revolution transforms education, commerce, entertainment, etc.

The greatest realization was probably the man's landing on the Moon.

Generally, from the beginning of the industrial age, but especially at the end of the past millennium, economical development, based on the spectacular evolution of Technologies, implied the consumption of natural resources. It seemed that the interferece between the man and his activities respectively the environnment produces. It is obvious that we all are more and more dependent on Technology.

Technological achievements are characterized by their big number, extension, appearance, sometimes disappearance, within short periods of time. All these caused contradictory human attitudes, some of acceptance and others of rejection, favourable or pessimistic prognosis. We can say that an essential relationship was born between the human species and Technology.

An elaborate study of these aspects is necessary and this paper, we hope, will be a contribution. In our opinion the problem has serious implications of an ontological nature. The development of Technology and of its products induced important changes with economic, social and cultural character for civilization, all reflected in man's consciousness and attitude. Answers regarding the accuracy of the development of Technology have to be given, in terms of an extending globalism and of some performant ways of study and communication, having some consequences with ethic and moral implications, concerning the participants in the technological act, their creation, including the artistic one, entertainment, even religious beliefs, anthropology, law and environmental protection.

The relationship between Technology and Science is also waiting for an answer, because there are points of view that try to give credit, to the idea that the purpose of Technology is not to find a truth but an artefact or a utility, which presents a physical materialization of some ideas from the origin. The point is that Science starts from ideas too. Both Science and Technology can have the same ideas as starting points, that developed, will allow different degrees of independence and interdependence between them. We specify this because there are other problematic ideas, for example the one that claims Technology is only an application of Science or even that both identify under the same term, *Technoscience.* We consider that these aspects can be better elucidated by appealing in an interdisciplinary manner with history, physics, informatics, biology, law, arts, engineering, medicine, sociology, politology, management, linguistics, etc. There are several points of view (Drăgănescu, 1996), including that of the authors, that Technology itself is a science having general characteristics, similar to the other particular sciences, already consolidated.

We specify here that it is about the essence (Heidegger) of Technology and not about some of its particular subsystems. In this sense we consider that, in order to speak about Technology, it is necessary to answer the following questions: *why, for whom, for what, how, with what, with whom, with what result (physical, biological, spiritual, cultural, value, etc.) for the human civilization*? These questions are valid for every domain of human activity and existence. Obviously, the material aspect dominates.

Considering the above ideas, we think that the main aspects of the concept of Technology, outlined and correctly acknowledged, can contribute to the enunciation of some ideas; that they could give explanations for the generalization of the concept, and could find a finalization by the creation of an individual philosophy (Fellows 1995). In this way, this new philosophy can be subscribed to the actual, general philosophy, from which the particular philosophies develop, such as Science, Physics, Chemistry, Arts, Religion, Biology, etc.

A first talking point has a semantic attitude, imposing itself by the fact that, at present, in order to answer the questions shown above, regarding *Technology*, the term *Technique* is also used. The two terms have in common the Greek origin in the word *techne* which means *art*, *craft*, etc. The addition of *logos* (*saying*) to the variant *Technology* belongs to the American scientist Bigelow. It seems that nowadays *Technology* is the more accepted form. We consider that the term *Technique* answers only some of the questions above such as *what, how, with what,* and in this case *Technique* becomes as a subsystem of *Technology.*

History can offer substantial arguments in order to find some appreciations regarding the evolution of human preoccupations, especially in the domain synthesised by the Greek term *techne* and of its influence during history, upon social development, human civilization, and culture. We believe that we are not making a mistake by saying that one of the fundamental conclusions is that Technology has a primary character in the appearance and continuity of humanity. From a historical point of view, philosophy was, from its beginnings, relatively passive regarding this issue. In the past century, preoccupations of this kind started to appear and they materialized in discussions and with contradictory positions. Philosophical and idealistic currents that appeared claimed us to be the witnesses of a materialization of the spirit with reference to the informatic model of the spirit (Bunge and others). On the other hand it talks about the *artefacts that think*, an idea that seems to be fascinating for engineers but especially for a certain category of materialist philosophers. However, these aspects are imposing in contemporary philosophy. The contribution of philosophy and philosophers in the actual conditions of technological development, and especially of its impact upon existence, impose with urgency. These problems and their elucidation have an influence upon the general attitudes of those who are active in the domain of Technology or are in contact with it.

Generally speaking, Technology constitutes a system having particular technologies as principal subsystems in which, specific theories apply. Technology itself can be considered as a subsystem of a superior one, namely the social system, where it interfaces with other subsystems (economy, culture, etc.)

Engineering is that which offers the possibility to put across ideas that are also necessary as starting points in Technology. Engineering can contribute to the conceptual and to clarifications (Mitcham, 1994), aiming at different types of Technologies. Engineering has important contributions to the development of Technology and it is obvious that it should be studied under its fundamental aspects, separately, but related to the development of Technology. This imposes the idea to consider Engineering as a science with proper, consolidated, philosophy.

In the above considerations a few aspects are pointed out, that we consider to be paradigms of the concept of Technology. We consider that they can contribute to the creation of a science and even of a consolidated philosophy of the Technology. We enumerate a series of other paradigmes that impose preoccupation, engendering the necessity of some answers. Here: creation and risk in Technology; the relationship of Technology and Humanism; the need for a technological education especially in the condition of the knowledge society; Technology and globalism; the democratization of Technology; the influence of Technology upon civilization; answer to the question if Technology is a power, and what kind of power, or it has a neutral character ?; the attitude towards the technocratic concepts; the role of Technology in sustainable development; Technology and biodiversity; Nature as an example for creation in Technology; firm attitude regarding some conceptions such as *endism*; *the end of the history*; *the decline of the human species* etc., all related to the impetuous development of Technology

One aspect that needs clarification concerns the artefacts and their relationship with natural things. Aristotle was interested in this matter, the idea being revived by other thinkers. The famous metaphor offered by Marx is well known, by comparing human architecture that has its origins in the mind as a specific human function, with the beehive, realized by instinct.

It has been shown that in our world changes occur very quickly and humankind also changes. This fact caused the French philosopher M. Puech to suggest that man has become a new species which he called *homo sapiens technologicus* (Puech, 2008). He came to this conclusion in considering that *technologicus* man's cohabitation with his products wasn't sufficient. He must become *sapiens (sage).* The essence of his problem lays in the different attitude between the way in which the world populated with old objects is seen, in comparison with the vertiginous promises of a new world. This is another fundamental problem that imposes a firm, philosophical, vision. The French philosopher sees the solution to this situation by using the small steps method, at which we add the necessity of its approach and of the ones mentioned above. It is impossible that the conclusions of these debates, especially in our country where they have a sporadic character, don't lead to the necessary clarifications and contributions in order to constitute a philosophy of Technology, on the terms of the existence of many preoccupations of this nature on a world level.

The tumultuous development of technological systems and the aspects that accompany it, imply a preoccupation with offering man the necessary abilities in order to integrate in his new universe, in fact, a technological universe. The ability to manage modern Technology from a philosophical, political, economical, cultural and social point of view represents one of the major challenges of the modern man in this century and it is foreseeable in the following one. The rational attitude referring to the limits of technological progress, with application in the human life, will permit an objective and optimal selection of the solutions in Technology.

Regarding the above considerations, we believe that an essential role resides with the school. This fact was understood in many countries where, in the curricula, Technology, was introduced as speciality study. We mention too, the emergence of technological universities, the e-Learning extension as working modality, etc. Therefore, the assurance of an appropriate content, imposes, because in some cases the Technology discipline has a descriptive, informative character, obviously necessary, and a less formative one, by which aspects regarding technological creation should be studied thoroughly, including its ethics, the social, economical, political, etc., even philosophical implications. The teachers and educators should have a philosophical background that permits the formation of a general idea on the world. This will be helpful in the instruction and education of their disciples in the spirit of humanism with technological specifics. Therefore the effects of their activity, displayed by the modification of the Conscience-Technology relation, will grow for the former.

According to the above, we point out, that the role of culture cannot be neglected, and especially the relation between the general and the technological culture, with a considerable weight in the first. This is in agreement with the *UNESCO* recommendation (in the '60s of the 20th century) of the enlargement of the general culture concept to the technological one, appreciated as being the most recent and important component.

In the sense of what we presented so far, we propose a new proper concept, adequate with the denomination of *Technosophia*, that we believe, it can evidentiate the philosophical aspects regarding Technology.

Being a supporter of the idea that philosophy isn't limited to the interpretation of the world, but that it also searches to find solutions to change it according to human necessities, we consider that *Technosophia* will acquire a larger sense and a more humanistic one.

TOWARDS AN ENVIRONMENTAL SCIENCE CENTRED ON INFORMATIONAL PROCESSES: ARTEFACTS, INFORMATION DYNAMICS, AND THE ANALOGUE-TO-DIGITAL TRANSITION

CRISTIAN SUTEANU

1. Bringing forward the Topic: a Woodcarver's Approach

It will probably come as no surprise if informational processes are soon recognized as some of the most important ones–if not *the* most important ones–to be considered in relation to our environment. We will argue in this chapter that the sooner this happens, and the more general this recognition is, the better.

Such statements only make sense, however, if their terms are understood unambiguously. This is certainly not the case here yet: key terms such as "informational processes" and "environment" enjoy such a broad spectrum of meanings that one could justifiably imagine a wide variety of points of view which would correspond to these opening remarks.

To delineate our approach to the topic, it might be worthwhile to follow the woodcarver's example and remove those paths or meanings that will not be applied in this endeavour. Branches of such an initial trunk could have addressed problems of environmental communication (Depoe 2006), public relations and environmental ideologies (Corbett 2006), philosophical perspectives on the environment (Foltz and Frodeman 2004), or even pragmatic aspects of crises management (Fearn-Banks 2007). We have chosen not to pursue these directions here; however, this paper will suggest that it is reasonable to foresee implications of the informational processes discussed here regarding some or all of the topics mentioned above.

An ontology designed to reveal the essential role of information in its multidimensional forms may still have to be written. The present paper will not attempt to outline this kind of ontology. It will not even speculate about the shape it might take. The goal of the paper is to propose some preliminaries to such a future possible ontology.

Our approach refers to information and its relation to the environment. We will not focus here on subjects such as information interpretation and the dynamics of meanings. On the contrary: we will attempt to move away from an anthropocentric point of view, and consider mainly informational implications of various processes from the point of view of their relation to change. We are more interested here in the role of information in relation to the physical world (Draganescu 1979, Landauer 1996, Duncan and Semura 2004); on the other hand, we will explore informational implications concerning different levels of organization beyond those of physical processes.

The paper is structured as follows. The remainder of this section addresses challenges related to the concept of information and presents some choices we have made, in this regard, in view of our investigation. Section 2 discusses the context in which the analogue-to-digital transition has taken place and explains why this transformation is critical from an informational point of view. Section 3 explores some implications of this transformation, with special emphasis on impacts regarding environmental change and human interactions with the environment. Section 4 is dedicated to discussions and conclusions.

We represent the view according to which information is part of physical reality, in which it has an important role to play. There are different ways in which scholars express the idea that information has a physical meaning. For Rolf Landauer (1996), information and its transformations are closely related to thermodynamic processes: he famously showed that information erasure has implications for entropy increase. Max Boisot (1995, 164) also finds that "information has a physical basis"; interestingly, however, for him the physical character of information is due to mediation of organisms that use information in their interaction with the physical world. While Kofler and Zeilinger (2006, 12) contend that "reality and information are two sides of the same coin", in this endeavour we consider information to be a component of "reality" along with–although different from–matter and energy (Draganescu 1979).

It is well known that the spectrum of meanings attached to the concept of information is enormous. Notions such as "data", "message", "uncertainty", "knowledge", etc. are used in relation to information in many different ways. We have discussed some aspects of the multiplicity and diversity of information definitions in another paper (Suteanu 2007, 69); a detailed examination of these issues can be found, for instance, in von Bayer (2004, 18).

The wide variety of definitions provided for "information" does not necessarily reflect conceptual immaturity regarding the crystallization of a concept. Actually, a source of confusion may be our tendency to use the same word, information, when referring to different features or processes. What we probably need most is clarification regarding the distinct informational aspects we wish to approach, rather than a unique and "complete" definition of information. In fact, while contributing to the early development phase of information science, Claude Shannon expressed the opinion that "it is hardly to be suspected that a single concept of information would satisfactorily account for the numerous possible applications of this general field" (Shannon 1953, 105). More recently, Mokros and Ruben (1991, 378) distinguished three different "orders of information", from environmental data to individual internal representation and finally to socially shared and negotiated knowledge. Their conception looks fruitful especially because they see these categories as "dynamic and always mutually defining". Generalizing this idea, one may expect to find a series of distinct meanings of information corresponding to different "levels" of processes observed in the material world, and, most importantly, these domains would not be encapsulated and frozen, but would interact with each other. On the other hand, certain common threads pervading information in all its forms should still be expected to exist, although "a unitary view of information as both a sociological *and* a physical phenomenon is still lacking" (Boisot 1995, 163).

In this paper we will attempt to move the emphasis from the vaguely defined concept of information to a notion that is simpler and easier to be circumscribed: information flux (Suteanu 2007, 70). We start from the premise that changes in the state of a system and the generation of new entities always correspond to information that specifies what has changed and in what way, what is new. The relation of information to change has been highlighted in different ways. For example, in his "Steps to an ecology of mind", Bateson (1972, xxv) defines information as "that which changes us". According to von Bayer (2004, 25) information "connotes change".

We define "information flux" as *the specification of change*. The smaller the change, the less dense (the more diluted) the information flux (Suteanu 2007, 70).

2. Paths within an Information Flux Economy

What would we do, if we wanted to generate artefacts, in ever larger numbers, based on limited information flux? Different strategies for "saving" information flux are conceivable. We will consider below three of them: *copy-and-repeat,* the *digital transition,* and the *virtual transition.* They are interdependent, and the digital transition plays an essential role in this context.

1. *Copy-and-repeat.* Creating copies of an existing feature requires a minimum amount of novelty: in theory, no novelty at all. One might thus create an endless number of identical objects without involving any information flux. In the material world, however, perfect copies are not possible. Therefore, an information flux (even if a diluted one) would correspond to the creation of any imperfect copy. There is a remarkable exception to this seemingly immutable law: while indiscernible features cannot be found on the macroscopic scale, aspects of identicalness and indiscernibility do exist in the quantum world. Not only are quantum particles identical, but they could not even be "marked" in view of future identification, as one can do in the macroscopic world. We are facing in this case genuine and unalterable identicalness. The quantum domain shares this possible presence of identical features with the world of information; for example, multiple instances of the same number cannot be distinguished from each other (even if their representations can be told apart, when they enter a process of computation it is impossible and meaningless to say which is which). Beyond the microscopic scale, however, one reaches the more familiar realm of the "different" and "unrepeatable".

In spite of the impossibility to generate rigorously identical entities, the procedure of creating copies opens the gate to massive production based on low information flux. Nature offers innumerable examples of such applications, when the same or similar processes lead to similar entities, from geometric modules in the mineral realm to individuals corresponding to the most complex forms of life.

Human society has also applied this approach. Mass production is a striking example, although the creation of numerous copies of the same objects characterizes even the earliest traces of civilization. At present, copies are being produced in an ever increasing number. On the other hand, copies are also more and more similar to each other: technology implements the repetition process of the "same" procedures increasingly well. Repetition of process and repetition of product go hand in hand and enhance each other. Standardization is a powerful form of ensuring rigorous repetition; it has been successfully applied to a variety of domains, from industry to the environment, from products to processes: an example is the ISO14000 series in environmental management (Mitchell 2002, 174). Globalization also adds strong momentum to this direction: innumerable similar products and services are offered almost everywhere on the planet (Suteanu 2007, 73). This is true not only for material products, from pencils to cars, but also for cultural products, such as music, books, and movies.

Population growth, an increase in the number of produced goods, and the dilution of information flux all occur at the same time, and they are not independent from each other.

It is intriguing to notice that the beginnings of significant mass production are directly related to information: the book was, as McLuhan points out, "the first mass-produced commodity" (McLuhan 1964, 174) and it became "the archetype of all subsequent mechanization".

The application of the copy-and-repeat principle in the case of books provided the starting conditions for a positive feedback loop with respect to information flux. Positive feedback loops almost always have critical consequences, since the strengthening of the causative factors as a consequence of their growing effects leads to ever-growing implications. This produces a variety of possible outcomes, depending on the actual system: most importantly, however, the accelerating effects that usually push the system out of balance cannot be stopped without an intervention that is external to the feedback loop at work. In the example mentioned above, the mass-production of books does not mean only that "identical" objects reach more and more individuals. Being increasingly exposed to common thinking patterns, individuals tend to produce similar patterns in their turn: some material, some not. The latter propagate identical thinking patterns in their turn, based on subsequently written books. This loop is supported by a system designed to shape thinking, starting with the youngest minds. The effectiveness of this type of dynamics is enhanced by the tendencies to advance towards a worldwide, increasingly uniform, education system.

If structures are copied and processes are repeated, what other means of information flux would there be available? One possibility is to run the same processes in a synchronized way. This requires a lower information flux, like a choir that needs less conducting effort if all the singers perform together rather than in small groups at different times. Our currently growing "shared commitment to abide by clock times" (Levine 1997, 95) is acting consistently in this direction.

Copy-and-repeat processes certainly have major and multi-faceted advantages. We are not discussing here the implications of these mechanisms in terms of production costs, material abundance, individual growth, health, access to learning, to the job market, etc.: we only point out their significance from the point of view of information flux.

If humans are so successful at generating quasi-identical features, from socks to windows to airplanes, it might be interesting to consider, even if briefly, what mechanisms seem to be at work in the natural environment. There are also pervasive tendencies concerning information flux saving at work there. Beyond the generation of similar entities, another, quite different mechanism can be distinguished; it relies on self-affinity. Self-affine features are characterized by the fact that their parts on different scales are similar with each other (Mandelbrot 1985, 257). The best-known examples are perhaps those of branching structures, such as trees or river networks, in which case the same kind of "pattern" can be seen when one advances from their smallest parts to larger and larger features; clouds offer another familiar example: one can see smaller and smaller parts that are similar to each other and to the whole cloud. The success of the science of fractals, founded by Mandelbrot (1975), mainly relies on the identification and quantitative characterization of scaling properties of real-world objects. For this discussion, however, it is particularly important that all these objects share a special feature: the fact that the same (or almost the same) pattern, the same "image" or the same "information" or at least the same statistically defined properties can be found over and over again, on each scale. This led to an information-based definition of such features: "A fractal is a mathematical construct that enjoys information conservation properties with respect to scale change" (Suteanu 2000, 5). From the point of view of the concept introduced above, this means that one encounters the same amount of change–therefore the same information flux–every time one moves from one scale to another. There is information flux conservation in scale space. For real-world features, self-affinity properties are always limited to certain scale intervals. Self-affinity represents an information-flux saving mechanism, and scale space offers a valuable way of evaluating such properties.

Self-affinity differs from the copy-and-repeat mechanism adopted by humans. It provides variety, as opposed to uniformity, while saving information flux. An impressive ensemble of features expressed both in space and in time (from landscapes to meteorological phenomena, for example) create a natural environment that always offers "diversity"; this plays an important role, not only for human perception, but also for the way natural processes are interconnected (Suteanu 2005, 133).

2. *The digital transition.* It is in this context that the analogue-to-digital transition can be seen as a key transformation occurring on our planet. As pointed out above, true identicalness and therefore zero information flux cannot be materialized. Beyond the scales of the quantum realm, all entities are marked by differences; they are always discernible. What would it mean to live in a world in which this law is violated? This question tends to sound like the starting point for a science-fiction exercise. However, we may not be as far from this situation as it seems, since, as we pointed out above, the realm of information can satisfy identicalness requirements. The realization of this contrast between the world of matter and the world of ideas has turned into a key point that supports the fulcrum of technology. Only an entity that has a limited number of accessible states could possibly challenge the barrier of identicalness; therefore, the digital form has been identified as the gate to identicalness. A large and fast-growing stream of applications has started to flow through this gate. The list of examples is endless. Converted to an electronic form and based on digital technology, entities such as texts, images, songs, etc. are carried into a world in which truly identical copies are possible. The digital forms of the copies are literally indiscernible from each other. One may object that all the forms transferred to the digital realm had belonged to the sphere of information in the first place: otherwise the transition would not have been possible. This is certainly true. However, before the digital transition, their materialization relied on techniques that allowed the identification of distinct copies. On vinyl or tapes there were unavoidable differences between the representations of one and the same piece. While the recording and multiplication of individual pieces clashes, by principle, with the uniqueness of each performance, digitization is pushing the information flux saving tendency beyond the limits of any analogue devices. Technology has not only made this possible, but has also produced affordable means to accomplish this task. The number of identical copies is growing explosively. While the quantity of information handled nowadays is enormous compared even to the recent past (Suteanu 2007, 71), a large part of the information flux is mere repetition; even identical repetition. Probably no other better forms of information flux dilution than digitization have ever been experienced by humanity.

3. *Shifting to the virtual.* A remarkable fact about the analogue-to-digital transition is its synchronization with the tendency for more and more processes to be transferred to the virtual realm. The massive, multi-level shift to the virtual domain (Suteanu 2007, 73) affects most aspects of

human life. It is not possible even to enumerate here all the processes that increasingly occur in the digital world. New aspects are being added continuously. Digitization and virtualization enhance each other. Business, entertainment, even education happen more and more in the world where the law of forbidden identicalness is violated. Spatial and temporal boundaries are changing fast and human activities flow towards a network-based virtual-realm-dominated world (Urry 2003).

Perfection in the act of copying can be achieved now in real copies, since they are based on the (material) representation of numbers; the size-limited rows of zeroes and ones can be reproduced with absolute fidelity. Can one thus speak about the achievement of an age-old ideal, the one of reaching the space of absolute mathematical purity–and not only as a lone thinker, but as a member of a community that has chosen to switch realms? It is precisely at the level of the community that the nature of these changes has been found to lead to profound alienation, according to Bugeja (2005). When one looks beyond the enthusiasm-inspiring approach to "digital perfection", one cannot help asking if, as a society, we are prepared to face such a transformation in a way that would bring human life to enhanced balance and fulfilment. This question would require an in-depth analysis that we are not going to open here. In the following section we will explore some implications of the analogue-to-digital transformation, with an emphasis on human-environment relations.

3. Informational Recycling; Composers and Decomposers

The matter and energy aspects considered nowadays from an environmental point of view include mainly minimizing a) the consumption of energy and matter, and b) the production of waste: in an attempt to decrease the human influence on the planet, one tries to decrease system inputs and outputs. Subsequent waste convertibility is also important: one would like to see the waste changed to some form of resource, or at least returned to the environment in a way that would make it indistinguishable from "natural" features. The most obvious informational aspects of these processes are those related directly to material transformations. Mass and energy conservation laws show that we do not, we cannot, "consume" matter or energy; Nicholas Georgescu-Roegen (1971) identified entropy as a key aspect of our interaction with the environment. In a general sense, we affect energy and matter by *changing* their forms, their location, etc. The nature of our impact is essentially informational. From this point of view, environmental problems can be seen as fundamentally informational problems. On the other hand, every action we take or do not take with respect to the environment crucially depends on information-centred, or at least information-dependent, processes, which leads to the recognition of the pivotal role of environmental information management.

From the above-mentioned perspective on information flux, a reduction in resource use involves the production of less change, and therefore less dense information flux. Integrating the waste back into the natural environment also contributes to the diminution of change. Recycling may dramatically diminish the amplitude of change (e.g. paper production starting from used paper rather than from trees makes a big difference from the point of view of processes of change).

However, recycling does not have to refer to material objects alone. Many types of systems are increasingly located in the virtual realm (or have their core values there). For example, generating and managing an organization involves a large amount of effort which is largely informational, even if it may also, and usually does, concern material components. When an organization must disappear (for instance because it does not work well in a changed environment), many different parts of that system may be worth preserving, "recycling", such as certain relations among its components, relationships that it helped create (involving individuals and/or other organizations), ideas that powered its modules, or even whole modules of the system. While material products are being designed with the eventual recycling processes in mind, one may also conceive organizations in a way that would allow flexible modules and relations to be reconfigured with ease. The same idea applies to projects, methodologies, and many other information-based "products". Decisions based on information flux economy criteria can have a direct impact on matter and energy consumption, as well as on aspects of time management.

While it seems practical to consider informational recycling in an information-dominated society, problems of a novel nature emerge precisely with respect to informational "waste" and "decomposing". McLuhan (1964, 175) had the intuition of the non-degradability of information-related entities when he said that "people began to act as though immortality were inherent in the magic repeatability and extensions of print". However, clear implications of non-degradability became visible only later. In fact, for material waste one can specify the approximate decomposition time (e.g. days or millions of years), as a function of substance and storage conditions; in contrast, informational products can last without limit: their material support may change, but the content has no decomposition limit. Information will be there even if considered "obsolete". The digital nature of information plays a key role here: for instance, if the information were transmitted using older, analogue types of support, absolutely identical copies would not be achievable; transmission imperfections would slowly lead to more and more information degradation. However, the digital character of the copies stands now as a guarantee for their identicalness–and therefore for the infinite life of informational products.

It is almost impossible to imagine a planet without the action of decomposers, which break down the complex compounds in dead bodies and bring the simpler substances back to the inorganic realm, to the soil and the air: in the absence of these bacteria, the planet would be covered with layers and layers of dead bodies… Actually, this seems to be similar to what we are facing in the informational environment. Supporting evidence for the existence of such a situation can be found in many fields of study. For example, while analyzing the fate of music that is transferred to the digital domain, Connor (2001, 473) finds that "digital technology abolishes degradation"; not just composers are immortal, but also interpreters, whose performances used to be attached to a fleeting moment. Moreover, "the river of time can start to swirl back upstream" due to digital restoration techniques that regenerate old recordings; from now on, they can be present, virtually, in perpetuity (Connor 2001, 474): "a voice that can sound forever, which will never decay or be lost because it can always be restored, threatens to turn into a kind of nondegradable debris." A similar process is identified by Roff, who discusses the status of old, fragile manuscripts transferred to digital copies (Roff 2005, 52). Such fast advancing accumulations occur in spite of the shift to "dynamic virtual concepts" made possible by digital technology in many fields, like archival science for instance (Cook 2001, 4).

It may seem strange to discuss technological advances of the digital era from such a point of view, given the advantages they provide to society. However, the immortality or non-degradability of products coupled with the unprecedented rhythm of their production could lead to problems for which we probably do not perceive all the ramifications and consequences yet. While the nightmare image of a planet covered by thick layers of non-degradable bodies would involve an "orderly" accumulation in horizontal strata, informational debris is different. It accumulates in a complex information space. A space in which we are learning how to navigate, a space that changes its configuration fast, and, most importantly, a space on which we increasingly rely for our success, and for our survival. It becomes clear that we are facing a race between the rhythm of growth of potentially-never-degradable products and the invention of orientation and navigation techniques, which are meant to help us move and avoid being trapped in our own debris.

On the other hand, loosely floating entities are harder to manage than coherent structures based on consistent relationships: flights to outer space are increasingly affected by the debris we have produced, which are now gravitating in huge numbers around our planet. Similarly, the pervasive fragmentation phenomena which are at work on every level of our interaction with information (Bohm 1980) are adding to the already growing challenges affecting our navigation in information space.

4. Discussion and Conclusions

Our efforts dedicated to a healthy environment are confronted with specific challenges when it comes to the informational realm, especially due to the analogue-to-digital transition. In contrast to material entities, informational products seem harmless, since they appear to leave both sides of the input-output equation (resources and waste) basically untouched. However, it is precisely this balance that is profoundly affected in the digital era. Non-degradable, perpetually existing informational entities accumulate fast, and the numerous paths through which they multiply do not seem to be compensated, which leads to some already visible–and many other foreseeable–challenges.

At first sight, this does not pose an acute danger, since there may be no obvious symptoms of "choking" in the informational environment: the fast and accelerated accumulation processes still seems to be harnessed using better and better tools. The need for new hardware solutions able to keep up with the growing requests, the demand for increasingly sophisticated data mining techniques capable of searching in more complex environments, may be just normal manifestations of progress. However, our confidence regarding the possibilities to generate an unlimited accumulation of products and waste in information space is reminiscent of the attitude we had not long ago, when we thought that oceans were able to absorb any quantities of pollutant. Here, the 'pollutant" that inhibits the finding of our targets in the waters of informational oceans is also informational. It might become increasingly difficult to handle such problems in multidimensional oceans, and one does not need to be exceedingly bold to imagine that new and unforeseen aspects of informational challenges will also emerge. What are the alternatives in this situation?

The informational recycling processes mentioned above would certainly decrease the contribution to new informational products. Supporting informational recycling, however, could easily turn into discouraging creativity: which of these would be more harmful? And who will be entitled to answer this question and make decisions? Other theoretically possible methods to combat the danger of debris-induced paralysis can also be found. An apparently simple solution would be to eliminate on purpose a part of the informational products. Actually, there seems to be no reason to keep all the informational products materialized in a way that would make them available; one could destroy some of them. Difficult questions would arise again with respect to those who would have the rights to decide what should be destroyed, and when. Moreover, much like one is reluctant to destroy computer files without preserving copies, one may also hesitate to destroy information products "forever": although no material entity would last forever, one gets quickly attached to the possibility to preserve something in perpetuity. The magic of the concept of perpetuity would be hard to oppose. A compromise solution could consist of the removal and isolation, rather than destruction, of some information entities: this would help to clear the space from information debris, and leave information still available to those who have encapsulated it. However, much like containers of nuclear waste, information capsules may leak, or they may be found and opened, and their contents would be released into the informational environment; the storage of informational debris could never be absolutely safe. Of course, questions would rise again: who is entitled to decide what should be destroyed, what should be "isolated", and who should have access to the isolation capsules? The problem requires our attention in spite of its apparently elusive character: given both the acceleration in information output and the growing importance of the information realm, the later the problem will be addressed, the more difficult it might be to handle it effectively.

One may thus consider in a new light the fact that information flux saving mechanisms are already in place; they are affecting natural processes as well as human activities. Information flux "dilution" (Suteanu 2007, 72) may play a key role in this context. Scale criteria prove to be relevant both for identifying domains of distinct informational behaviour (quantum vs. macroscopic realms) and for recognizing categories of mechanisms (self-affinity, identicalness, or other manifestations of symmetry).

Supporting the emergence of an environmental science that would focus on informational processes, only a small sample of which have been addressed here, promises to be an important, as well as a tremendously complex, endeavour. Perhaps the first step should be to include an information-centred approach to the tools we use in the exploration of the environment.

The informational aspects of our interaction with the environment are still less transparent than those referring to material entities. This does not make them less important: as Floridi (2007) points out, “information is becoming our ecosystem”.

PHILOSOPHICAL PROBLEMS OF QUANTIC PHYSICS IN MEASURING PROCESSES AND INSTRUMENTATIONS

DAN MILICI AND MARIANA MILICI

Introduction

Education implies two fundamental aspects, which will be given definition and understanding hereof.

Space or horizontal dimension (axis). Through this we nominate the extension of education in all living and social-human life areas. This interpretative perspective leads to eliminate the detrimental mentality that education involves only through scholastic institutions. The contemporary social evolutions revealed the existence of numerous other sources of education, extra- or non-scholastic; configurations besides the formal education (of the scholastic type). Two other great educational areas – non-formal education (achieved through numerous social institutions with educational relevance like cultural-scientifically institutions, mass-media, economical organization) and informal (which covers the wide areas of numerous spontaneous, unsystematic and transitory influences, which take place inside the space of social contacts of all types, in which the individual is implied at different moments of its extra-institutional being).

Temporal or vertical-longitudinal dimension (axis). Its major aim is the extension of education during the entire life of an individual. The new acquisitions from the domain of ages and learning psychology have demonstrated, undoubtedly, the persons' capability to learn and to be educated at any age, of course in ways and with intensities modelled by the psycho-social particularities at different ages.

The most durable and efficient innovations are those such as the beneficiary assimilated, that is those adopted because they satisfy a person's specific needs. Let's take into account that happens at the concrete level with transformations produced by the new technologies of information and communications:

Firstly, we observe the conversion of cultural content from all over the world into a digital form, thereby making the produces available to everybody, everywhere, and at any time. The wide area and high speed communication networks, connecting the computers from apartments or off student's desks to high capacity digital libraries, have changed the cultural conditions in which education developed. The loneliness is bygone and there is a substitution taking place of insufficiency and superannuation with its ampleness and sufficiency.

Secondly, multiple ways have been developed to represent information, to simulate interactions and to express ideas, extending the acquisitions of intelligence, thus altering the spectrum of civilization, modifying the requirements of participation in the culture. The epistemological developing acquires interesting aspects. Thoughts relate closely to language, the formal symbolizing from mathematics and logic being seen as an extension of the various current linguistic aspects. The digital medium evidently extends the area, being used to acquire information and to express ideas in different ways – verbal, visual, or a combination of both. As a result, for educators it will be more and more difficult to favour the operating of verbal language in detriment of the other ways of expressing.

Thirdly, people externalize various current abilities – to compute, to write correctly, to memorize, to visualize, to compare, to select – regarding the digital instruments with which they work, thereby obtaining a real mastery concerning these abilities, sometime the results of education. The digital technologies amplify the individual potentials. The text processors, for example, warn in case of any word written incorrectly or in case of grammatical errors, the worksheets allow everybody to quickly and correctly make calculations using extremely complex formulas, the data bases allow even those with low retention to manipulate entire information suites. Other various forms of professional working instruments reduce the ability levels needed to participate effectively in a large area of cultural activities.

The new paradigm is bound up with the effects of classical physics encompassing:

1. technological development;
2. new physical theories;
3. dealing with reasonabilities for education.

The different fields in which the measurement techniques can be included also partially covers the sphere of the automation systems. However, in the large majority of the applications, it is man that receives the information that results after the measurement. However, much they have endeavoured, the researchers did not succeed in achieving; such as the complex automatic devices that one can say has its own thinking, sentiments or judgment. What does it mean, in fact, to think? What is meant by the mind? If the mind exists, how much it is connected with its material support which it is associated with? Should there be any thinking independent of such structures? If we analyze the situations, we observe that the notion of *artificial intelligence* is improperly assigned to the electronic systems so named (systems which work after a perfectible program but which do not understand how it is made). We did not know if it submitted to the *laws of physics,* and after all what are the *laws of physics?* Something else other than an *emanation* of the human mind based of course on logics and mathematics (but also on experiment)? Certainly how long will not exist *exact* answers at these questions, as well in the metrology field will be problems which will be unsolvable or partial solvable.

4. Although the metrology is constituted as an independent science, with explicit laws for the specialists who practice it, the scientists and the philosophers' who show interest in this field are very preoccupied by the impossible things. There are opinions which affirm that the science is possible in general because there are impossible things. The action of measurement has mostly the equilibrium (compatibility) on the base. There are equilibriums described mathematically by special solutions of the equations, whose stability was not decided. For this decision, impossibility has an impact over some problems of real interest in metrology; the equilibriums must imply the playing of a very great number of interactions.

5. What special things must we know about *nature* to appreciate correctly an achieved metrological action, and how must we achieve it to find the information we are interested in with satisfactory accuracy? If our conscious perceptions are not something else that are the result of some algorithms succeeding, as more adepts of *artificial intelligence* believe, then should not of great importance which are the nature laws. Does the experimenter's mind influence the experiment or the measurement result? Quantum physics affirms that the measurement modifies the measured medium and influences it with intensity probabilistic computed, therefore the measurement result is different to what it would be if the measurement was not carried out. Moreover, the incertitude principle from quantic forbids us to determine with accuracy both the position and the impulse, therefore, as precise as we should be on measurements, only a half of the information can be obtained in this case. Does science need to go in search

of new and more accurate devices and methods? Is the result of one measurement the value found or the value created by our mind?

Philosophical Problems in Measuring Processes

Beginning from the history of engineering performance forecasting technology, men were always fascinated by the possible evolution of precision and of the performance marginal limit. In the past, people did not even take notice of progressive evolution. Now it is difficult not to notice it. Every adult, during his own lifetime, has been able to observe the appearance of personal computers, the Internet, and cell phones, while practically anyone who wants to do so is able to own his own automobile, and so on.

Sociologist M. Sukharev, in his work, "The Explosion of Complexity," presents the sequential steps of these changes:

> There is one more pattern that is evident in societal development - the acceleration of growth of complexity with time. For tens of thousands of years tribes lived on Earth armed only with spears or bows and arrows. In several hundred years we have leapt through industrial technological civilization. It is unclear as to how many years the computer era will last but the current speed of social evolution is unprecedented. (Sukharev.)

In this article, on the basis of analysis of extensive scientific and factual materials, a hypothesis will be formulated providing answers to the noted questions and allowing predictions to be made for both the immediate future of the evolution of precision measurements as well as the general direction of evolution in the far distant future.

For some, chaos and catastrophe is: a science of process; a new way of thin-king; a return to the study of phenomena on a human scale. For others, the essence of chaos theory is the mathematical model used to formulate "strange attractors", "fractals", and "phase space." In our attempt to integrate chaos theory into the field of measurement performances, both of these doctrines will be examined.

Forecasting the evolution of human development is a modality of knowledge, focused on elaborating predictions based on development trends.

We can say that measurement has emerged with human thinking. What we cannot say for sure is how it emerged: out of necessity, gradually, or as a revelation. The fact is that life is generally closely related to the act of measurement, an act that came to be a measure of human civilization. Both in daily life and especially in industry, various designs of measurement and control devices have been obtained (especially in recent decades), impossible to assess the number.

Metrology is important both for science and for the economy because only by metrology we can express physical quantities and their values, and improve the experimental methods underlying all physical phenomena research.

*

If there is a *philosophy of metrology*, one of the most important factors that it underlies is the equivalence between devices and systems when working. Meter is seen now as something much less important, allowing for the algorithm, the program as the main player (recently more indirect measurement techniques have been developed). It seems still that there are other important factors that influence the measurement process and come from physics, technology and logics.

An essential property of a formal mathematical system used in metrology is that, for a statement expressed as a given mathematical data, to decide whether or not a string of symbols is a demonstration in the system, and should be a calculation problem. The essence of demonstration concept formalization consists in that there not be made other trials on what a valid argument means.

It proves that in any formal system there is always a correct algorithm for finding a demonstration if it is possible, because it supposes that the system is formulated in terms of a symbolic language (language that can be expressed in terms of a finite *alphabet* of symbols).

If Hilbert should be able to build his mathematical system (a system of rules and axioms strong enough to decide, using formal demonstration, the truth or falsity of any mathematical statement expressed correctly in the system) then it would be a general algorithmic method to decide the truth or falsity of the statement in the above.

The reasons given above, regarding the limitations of algorithms, have been expressed *in principle*. It was not discussed whether the algorithms can also have practical applications. Even in case of some problems for which it is clear that there are algorithms, and how they can be built, it may be necessary for more inventiveness and work to transform them into something usable. These problems are related to techniques and understanding of phenomena. In recent years much of the work of research teams has been devoted to construction, understanding and improving the algorithms (as an area of research in expansion and rapid development).

As noted, even among mathematical problems that are of an algorithmic nature, there are classes of problems which are largely intrinsically more difficult to solve algorithmically than others. The theory dealing with problems of this type is called *complexity science*. Complexity theory is important because it opens another issue, other than that related to the question of whether the processes and phenomena are algorithmic, namely: whether the processes that are known algorithms are algorithmic in usable mode.

Similar problems can also be read from the theory of relativity. The world we live in is governed by relativity because the speed of light is finite (we don't know why the speed of light has the value that is in the entire universe). If this speed were much smaller, slower moving objects would have distortions of space and time when their speed was closer to the speed of light, less energy would be available if the matter could be annihilated in nuclear reactions, and hence the matter would be more stable. These aspects would completely reformulate the entire metrological legislation especially since the usual processes in metrology are not algorithmic in usable mode.

There is a three-way exchange between time, energy, and information, which is controlled by limits on the amount of information that can be achieved with a given energy budget (if we make a determination slowly, we use less energy than if we do it quickly). It follows that the reversible processes are more efficient (wasting less heat) than the irreversible. It sometimes uses excessive amounts of energy for technological or informational progress. The resources used for these purposes should be increased in line with the limits that nature imposes in terms of how we can provide fine time measurement; how much energy should be spent to acquire information; how close to the critical state complex systems we see are; and how our technology is sensitive to errors and random amplification of uncertainties.

In metrology, there are two distinct areas: the physical world containing materials and energy, and the mathematical world (imaginative, intuitive) used to decode the patterns found in the real world. In the measurement domain, an essential component is *communication*. Impossibility to use the information is clear even from the average level of a measuring instrument without such communication. Starting from the dictionary definition, disclosure, notification, presentation, communication is a process that, from the angle of communication science, has several basic components: transmitter, encoder, information, channel noise, receiver, and being ideal to be a bidirectional communication even if it does not use the same channel or same encoder. From this perspective, the measurement can be regarded as a complex system, the measurement characteristics being not the only ones that can characterize it.

Furthermore, the measurement process does not end with taking or receiving information. Information may have an effective influence on the system that receives it. The process is called the *communication effect* generating an *intentional* stimulus-response process (because not every stimulus-response process is a communicative situation). In this context, communication becomes a process by which a transmitter sends encoded information to the receiver through a channel in order to produce certain effects on the receptor.

In terms of information, functions performed by humans in relation to the device or the environment are:

- *Monitoring function* - closely related to the choice of reference systems. From the informational point of view the only reference system is related to the experiment and not by the used device. The monitoring function is not identified with that of perception because it involves additional aspects concerning the dynamics of processes such as attention, designation, recognition, mental model of the experiment, etc.
- *Diagnostic function* is that by which the operator is allowed to intervene directly in the experiment by the informational system. The conditions for successful solving of this action are the existence of a mental model of the entire managed process, and sequential training of some operational models based on received information.
- *Decision function* is the only way to be able to establish a continuity in the activity of the operator, it being required by the informational nature of the human itself and by its role in the energy-information system.
- *Adjustment function* that occurs in order to maintain the values of parameters within normal limits.
- *Control function* that appears as an equivalent of control function if we look at it in terms of actions interposed by the operator.

In metrology we are confronted with a complex combination of at least four general theories: a theory of formation and differentiation systems, a theory of evolution of these systems and concepts, a theory of creation of meaning and purposes, and a theory of media. The objects of these theories are considered interrelated in the formation of a global metrology system in the sense that technological developments lead to extended systems, more complex and differentiated and the systems develop ways of communication both specialized and characterized by a higher level of generalization, in order to overcome the differentiation level already achieved. A coordination of different subsystems is achieved consecutively with the media leading to a relocation in time and space.

One of the essential requirements of the modern theory of measurement systems requires these representations (measures) as opened for environment and unstable, that changed the fundamental problem of classical metrology that developed the system as a whole, composed of parts with the risk to define the identity ontologically. The greater difficulty arose when it was seen that the whole represented more than the sum of constituent parts. Even when the concept of device has been replaced by the concept of system, this still does not overcome the model of *determined* systems that could accept occasionally only one type of measure operation. At the changes of the environment in which it were placed, the systems could react in only one way, always predictable.

If we consider that to each point from the material space we can attach a value for temperature, then the densities field implies automatically a temperatures field and thus a field of internal energies that can be determined with precision. This would be only one of the applications that tend to provide new insights into the measurement of void and thus in widening the domain of metrology application.

The most important of our conscious sensations is that of time passing. *It seems* that we move from a very clear past to an uncertain future. As we feel it, the past ends and there is nothing to be done regarding it. It is unalterable. What we know about it may come from evidences, from memories, from deductions based on them, but we do not doubt the *reality* of the past. On the other hand, the future seems to be still undetermined. As awareness of the passage of time, the nearest part of this vast and unknown (apparently) future is carried out constantly as present reality and enters the unchanged past. Sometimes we feel that we were even personally responsible for what has influenced this choice, or we feel helpless spectators as, inexorably, the well determined past boundary makes a place in the uncertain future.

Physics is the opposite view. All equations in physics, equations correctly describing nature, are time symmetrical. These can be used equally well for a sense of time and for the opposite. Newton's Laws, Hamilton's equations, Maxwell's equations, Einstein's general relativity, Schrodinger's equation, all remain unchanged if we reverse the time direction. According to relativity theory *now* has no correspondent in reality. The closest representation of this concept is the so-called *simultaneous space* of the observer from the *time* space.

Conclusion

Theories in physics also have other important components (not only temporal evolution equations); some of these assuming the temporal non-symmetry. The most important of these is the Second Law of Thermodynamics (*entropy of an isolated system increases in time - or remains constant for reversible systems*). The concept of *entropy* relates only to the disorder manifested clearly in the system. For a system isolated from the rest of the universe, total entropy increases so that if the system starts with a particular clearly shown organization, it will erode over time, becoming a meaningless motion of particles. It seems that the second law of thermodynamics expresses a despair that there is an universal physical principle which states that any organized structure evolves towards destruction (faster or slower).

But can we expect as well to define the concept of *"disorder manifested clearly"* needed to make the concept of "entropy" being mathematically well defined? Could it be that it depends on the precision with which different observers could make measurements on the studied system? It could be that an observer, with more accurate measuring devices, would be able to obtain more precise information about the microscopic constituents of the system than others (thus assigning different values of entropy). It may be that aesthetic assessments of different observers are involved in what they consider to be *order* rather than *disorder*. Relative to these problems of subjectivity, it is remarkable that the concept of entropy is still useful, that it can give accurate descriptions of scientific value. The reason why it is useful is that the changes produced in a system in transition from disorder to order, expressed in terms of detailed positions and speeds of individual particles, are truly enormous and will fully cover any differences relative to what is or is not a *"shown order"* at the macroscopic state.

We presented here some aspects of theoretical physics that justify the complex problems arising in finding a coherent reasoning which models the thermal phenomena occurring at a temperature sensor, staring from the quantum level and ending in the macroscopic area in which we operate. The physicist sees that there are limits for precision and certainty, that there is a boundary even for the pure thinking of theoretical physics. A part of this boundary is the scientists themselves as thinkers. The laws of physics, which in fact underlie all instrumental determinations could be expressed in terms of a *complete* mathematical system, but almost always the engineer is concerned with ensuring himself that he obtained the *right* system rather than a *complete* system.

*

The building of one informational society (which will represent the passing to the *knowledge society*) cannot be achieved without research and investments in projects, either in the IT&C field or in the education field.

Fundamental research is an expression of human aspiration for world understanding; as well as a source of some human aspirations on what should become the world.

The aspirations are desires for which exists in principle the possibility of some concrete achievements. The aspirations are the first step towards the materialization of "something" that exists only in spirit and that can gradually become the expression of spirit manifestation in physical reality. There are *own aspirations,* and *aspirations* that have no sense unless they are common, if around them a sufficient number of individuals are united by the same vision. Only then a *functional project* is born.

The complexity of the issues and the difficulties that arose were overstepped due to the interdisciplinary nature which was at the base of the team's formation, each member being in a position to help by their own knowledge to the completion of the achieved studies.

The "model" and "modelling" notions reached deeply in the instrumentation theory and practice. The functions (cognitional, cultivation-educational, instrumental and normative), accomplished by the models used to solve the problems of the instrumentation theory and practice, can have various characteristics. The efficient management of the training process is coherent with the use of various models.

From the specialized literature it is found that the instrumentation forecasting problems centre round the use of the various forecasting methods (extrapolation method, modelling method, and examination method). The forecasting made using the extrapolation method allows the formation of measurements results on the hierarchical system, on the basis of the study of some adequate laws from the previous period. The forecasting accuracy can be right if the forecasting period is shorter and if the data are more.

TAGS AND FOLKSONOMIES AS ARTEFACTS OF MEANING

ALEXANDRE MONNIN

1. Introduction

The advent of the so-called Web 2.0 and the Semantic Web, for all their seemingly apparent (Floridi, 2009) differences, was instrumental in a renewed interest in questions that used to be addressed solely by the philosophy of language. Amongst these, the problem of meaning is paramount to many a Webservice. The fact that philosophical *problematiques* of such a magnitude as this one are brought up both in the design process and effective use of technical devices, forces us to realize that some topics are already moving from the traditional realm of theoretical questions to an area packed with innumerable "artefacts of meaning" - tagging systems among many (we will focus on these for the time being).

This shift, that we've chosen to dub "artefactualization" elsewhere (Monnin, 2009a), bespeaks the now common coupling between philosophical questions and technological realities. In other words, a process that exhibits continuity as well as radical change and deeply subverts established views about the relationships symbolic forms and artefacts entertain with one another.

*

In this paper, we try to give a characterization of tags, drawing from previous work accomplished in the wake of the Semantic Web "identity crisis" and at the same time with regards to keywords, descriptors, and other lexical components found in library sciences in order to dispel widespread confusion on these matters. Eventually we show that the meaning of a tag is not a question that can be restrained either to its technological or symbolic dimensions.

2. Tags: Material Props and Labels, between Access & Reference

2.1. The Three Canonical Dimensions and the Access/Reference Duality

In his initial definition of folksonomies, Thomas Vander Wal was keen to underline their dependence on tags, thus observing the latter's precedence:

> Folksonomy is the result of personal free tagging of information and objects (…) for one's own retrieval. The tagging is done in a social environment (usually [my emphasis] shared and open to others). Folksonomy is created from the act of tagging by the person consuming the information. (Vander Wal, 2007)

Indeed, two years had to pass after the creation of Muxway, the first self-proclaimed tagging system, for the coinage of the word "folksonomies" to happen. Since then, the lack of a proper characterization of tags has consistently remained conspicuous. Research eagerly followed the trail of folksonomies, understood as a collaborative phenomenon, in the hope of leveraging users' contributions on a large scale (also known as crowdsourcing), rather that foundational trends. It is also to be acknowledged that the topic remains quite laden with confusion mainly due to the constant equivocation between the recurring and holdall notion of *keyword*, *descriptors* pertaining to documentary languages, natural language *words* used to launch search engines' queries, HTML <meta> "tags" embedded in Webpages and, strictly speaking, tags. Hence, the existing vocabulary in library sciences no doubt acted as a serious hindrance which prevented any clear-sighted depiction of the main issues at stake in this debate for a relatively long time. To draw such a depiction with accuracy, it is necessary to focus our attention on the ever-changing technological context which saw the emergence of tagging systems. Only if this condition is fulfilled will it become possible to bring the "symbolic" dimension of tags to fore and show how inescapably it is intertwined with complementary technical aspects. *Symbolic* is here to be understood as a broadly encompassing term. We have to acknowledge the influence of Belgian philosopher of technology Gilbert Hottois' thought, which contrasts "symbolicity" and technology – yet we also try hard not to underestimate the symbolic activity of technosciences. For Hottois, it is on a level of its own as evidenced in this excerpt, strikingly reminiscent of the Semantic Web:

> Le langage des technosciences se veut purement objectif. En termes logiques, ce devrait être un langage purement extensionnel ou référentiel, dépourvu de toute étendue de sens. Un *langage qui étiquette le réel afin d'offrir des prises et de permettre l'organisation collective des relations opératoires* [my emphasis] – techniques, mathématiques – au réel. Un tel étiquetage est conventionnel et instrumental : ses mots n'apportent rien – ni sens ni supplément de maîtrise – à la maîtrise technomathématique des objets, des opérations et des processus qu'ils désignent. (Hottois, 1996, 91)

We thus go as far as to temporarily subsume, under the word « symbolic », iconic representations as well as linguistic signifiers.

While no canonical definition of tags is available on the academic market, thanks to early (Newman, 2005a) attempts to model tags with such ontological web languages as RDF and OWL, tag ontologies do however exist. The reifications they afford will temporarily palliate the lack of genuine definitions and provide this investigation with a commonly shared starting point – albeit an ontological one which may sound a bit ironic considering how much publicity the alleged opposition between ontologies and folksonomies has received (Shirky). (Wikipedia gives the clearest imaginable definition of reification: "making a data model for a previously abstract concept".)

Newman's tag ontology (the same is also true of Thomas Gruber's *TagOntology* from 2005) aims to describe a single tagging process by distinguishing between:

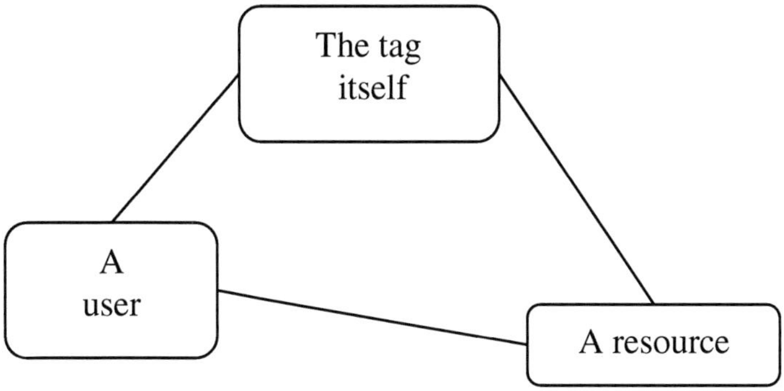

Fig. 1. – The three axis of Newman's tag reification

For all its limitations, a quick survey would nevertheless show that such a threefold division is barely challenged by any alternative. With the ensuing risk of obfuscating the dual nature of tags: as much *material* devices (an echo of the concrete nature of the object after which they were named) as *symbolic* ones. To conflate the two aspects is tantamount to overlooking the simple fact that *symbolic bonds between words and things do not in the least require to be technically implemented in any way.* No technical apparatus is necessary for a word to point to an object, no artefact will ever make up for this possibility; in other words, reference pertains to the domain of semantics (or pragmatics, thereof - though the question remains open as to whether or not others factors can affect it – cf. *infra*). On the other hand, every tagging system is implemented according to the rules dictated by the needs of the website it serves (who's allowed to tag? what? how? etc.). *Reference* is thus *de facto* complemented on another level by the association, on the technical side of things (the Internet being a physical network were information is exchanged, this should not come out as a surprise), of a *tag* with a *resource*. (The word "resource" is here used in the technical sense it has received from the W3C. A resource thus understood can be anything. Cf. *infra* §3.3.)

Its ins and outs concern interfaces informational design and the tangible realities of networks and protocols - especially the Web architecture centred around the HTTP protocol (RFC 2616, 1999) and URIs (RFC 3986, 2005). "Access" is the name Patrick Hayes (2006) has given to this relationship in the ongoing debate about the "identity crisis" faced by the semantic Web (Halpin, 2006). He convincingly underlines the necessity to carefully dissociate the two dimensions in his discussion of URIs. Our main assumption is that it is essential to reckon what this discussion revealed, and transpose its result to our endeavour to characterize tags in a twofold fashion; first, as words, or rather potentially meaningful strings of characters; then as a « material » reality granting access to a resource and tightly constrained through limitations attributable to the computerized systems it belongs to (be it the Web or a local application like *Photo Gallery* on Microsoft Vista OS – cf. *infra*).

Deprived of this distinction, Newman's endeavour raises serious issues. First and foremost, such an ontology's worth has to be evaluated with regards to what it makes more explicit as much as what it obfuscates. Any *description* of tags revolving around the same three axes without showing any concern to the complexities of interfaces would seem at best questionable. Yet, once implemented, it will provide an "adequate" means to achieve (or rather *prescribe*) interoperability between miscellaneous tagging Webservices, thus giving birth to a unified conception of tags without any room left for second thought.

Computer ontologies serve not only to describe reality but rather as tools to *build* knowledge systems around a given set of *representations*. Belonging to the same technical realm as the entities they model – at least in the digital context of the current discussion, one simply cannot confuse them with "mere words" wholly deprived of technological efficiency; a good reason to show special care and dedication towards the objects being modelled.

2.2. From Tags to Machine-Tags: Blank Spaces, Labels and Material Tags

Access relationships differ from reference inasmuch as the former are altogether causal in nature, therefore material, and relate users and resources through a relation of *ostension*. Ostension as philosophers have noticed, in particular since Wittgenstein (1953), inescapably points to its objects in an indeterminate fashion, or, to update our description in a way that better suits the current realities behind the Web, to the resource it designates.

It could be argued that causality is (obviously) central to causal theories of reference, which would undermine the distinction we're trying to draw. However, whereas in causal theories ostension plays a major role, they also tend to rely on naming conventions reintroducing, as a result, an element that escapes causal reduction. As Hayes puts it:

> "reference to accessible entities still differs from access. Establishing reference by ostension requires naming conventions". (Hayes, 2006)

The label itself is nothing more than the string of characters inscribed on the tag – construed as a material support and artefact belonging to an informational network. While contriving access to a resource (informational or not, cf. *infra*), this prop also allows users to add any required bit of texts. It thence becomes possible to index, evaluate, share or find again objects that previously overstepped the customary limits of annotation. This is precisely what the traditional post-it – a material device by all account – offers: a blank inscribable surface attachable on other (supposedly) non-inscribable surfaces. To put it bluntly: annotations where it used to be lacking. For instance, the software « Time Line », developed in France by the IRI (see *bibliography*), makes it possible to set apart and annotate movie sequences at will with tags (a feature also found on Yahoo's recent *VideoTagGame* (SANDBOX, see bibliography) with an emphasis on collaboration typical of GWAPs, *games with a purpose*).

Another manifestation of the wide dissemination of ad hoc apparatus supporting input of meta-data by users would be Window Vista *Gallery* (Smith, 2008, 197-201). The main feature it offers is the possibility to tag photos on a local machine not simply by tying together a tag and a digital resource but rather by injecting the tag into the said resource. Contrary to post-its, often used as mere reminders with no guarantee that they won't get lost themselves, *Gallery* keeps track of every tag in the system to secure lasting access to all marked resources. The rationale behind it is no longer akin to that of shortcuts for it is well-known that whenever the target of a shortcut is moved to another directory the shortcut will still point to the same direction and miss the mark. Incorporating tags inside resources, provided tags are traceable without fail, will make the former impossible to lose – just as tags themselves. Hence both tags and tagged resources happen to form a new whole, sharing the properties of its parts, including infallible traceability.

Nothing, as betokens the analogy with paper tags, constrains users to write down strings of characters that necessarily take the shape of words – let alone lemmatised ones (whatever the syntactic constrains each tagging system tries to enforce on its users; for instance tags may be separated by either by spaces or commas). It follows therefore that it is simply not desirable to roughly equate tags with any known linguistic form. The reality is infinitely more complex than the previous schema would let us assume, for the reason that the status of the "tag" entity in Newman's ontology isn't critically addressed. However, it is not so much advisable either to get rid of a threefold partition, which has done much to help operationalize tags by adding relevant handleable ontological attributes (the choice of every facet is left upon the user; see Monnin (2009c) for a brief historical reminder), than to expand it by introducing more complexity. Namely, the already mentioned distinction between the material reality pointing to a resource (or rather, the *representation of a resource*, as we shall see) thanks to their common belonging to a given technical network, and a semiotic and linguistic relationship carried out through items that eventually boil down to strings of characters.

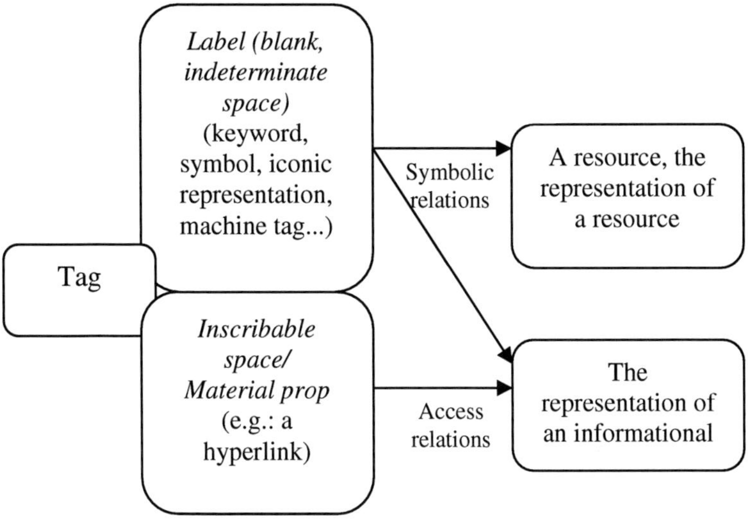

Fig. 2. – The twofold nature of tags.

Thanks to the freedom characteristic of what we chose to call an "inscribable space", it is even possible to mix these two dimensions. Actually, Flickr's machine tags do just that. As a lightweight way to bring more structure to users contents, they articulate three dimensions (actually this is where they first got their name: *triple tags*):

namespace:predicate=value

Breaking data following this pattern allows Flickr teams to send the "facets" (as in "facet classifications") of a machine tag to a database where they can be queried through the website API. In this way people are offered the opportunity to formulate any query starting from a given namespace, a predicate and a value or any number of combinations of the three.

These tags provide a borderline case which nevertheless illustrates the relevance of the access/reference duality integral to the model depicted in Figure 2. Historically, two possibilities make themselves available.

Either machine tags will be treated as mere *triple tags*. Triple tags used to be labels written according to the above schematization: their syntax heavily influenced by the XML family of languages, microformats and the spread of RDF graph-like models, provided users with an effective and low-cost way to express complex relationships with only one label. The latter, once embedded in a URL would then give rise to a tag that would in turn give access to any dereferenceable resource on the WWW.

Or (a possibility not incompatible with the previous one as we will see), they no longer need to be associated to a URL. Their parsable syntax made them sufficient to grant access to information resources by way of queries launched through APIs of websites like Flickr where triple tags, once used as mere symbolic tools, became machine-readable; that is, where triple tags can be used as machine tags (*thus bringing the access/reference duality at the very level of the label itself*).“For the moment, *machine tags* are principally an *API "thing"*" (FLICKR). Such a conclusion could even be generalized to any term that is part of a query made on a search engine.

Since no explicit rules require that labels should be limited to words or specific human-readable signs, nothing prevents using computer code either. Even more so since, as vague a syntax as the one people used simply as a means of abbreviation, gave birth to a new microformat on its own as soon as machines were programmed to “interpret” it the right way. Although extremely crude, from the proto-syntax it used to be, it nonetheless became *ipso facto* a legitimate computer syntax. From this angle, machine tags are the logical product of a tendency that leverages the fundamental duality of tags.

3. Which Semantics for Labels?

A striking feature in the debate surrounding the status of tags undoubtedly is how often they are put on a par with keywords without further ado. For that reason, it is a pressing necessity to give precise definitions of those notions borrowed from library sciences to which they are repeatedly assimilated. Otherwise, there would be a high risk of obfuscating every attempt to clarify questions related to the semantics of labels (and not, properly speaking, of tags).

Joshua Schacter’s comment on Wikipedia discussion page:

> “While keywords are not new, I believe that tagging is a larger concept than just assigning keywords to things, however - I feel that it also includes the retrieval of the set of used terms/keywords/whatever upon view of the items. Additionally, I am reasonably sure that I named this.” (Schacter)

Given their conventional nature, lexical entities in documentary languages are very well defined (at least compared to other lexical components like words whose nature may include normative elements but whose semantics is not the result of an explicit stipulation from end-to-end). Let's examine them.

3.1. Descriptors

International ISO norm 2788 for the establishment of monolingual thesauri, already quoted by Manuel Zacklad, defines them as

> "*vocabulary of a controlled indexing language formally organized so as to make explicit* a priori *relations between notions (for instance, the generic-specific relation).*" (Zacklad, 2007, p. 5)

The first consequence of this definition – if indexing is all about giving an account of a document's content – namely, the lack of referential relationship, provides indexers with the possibility to have recourse to an indefinite number of descriptors. These descriptors cannot therefore be in a one-to-one correspondence with subjects. The latter can however be apprehended through a bundle of notions drawn from a thesaurus.

Thesauri consist in networks of a limited number of relations between their components (descriptors), including genericity, specificity and association as regards the main ones. In order to pick up a descriptor, it is necessary to spell out, upstream, all the connections it entertains with other descriptors, for descriptors only exist as far as they're interwoven with each others in a controlled vocabulary which allocates a defining place to each and every one of them. Accordingly, the meaning of descriptors needs to be properly understood from an inferential point of view for relations like hyponymy, hyperonymy or synonymy only betoken the intrinsic order of the thesaurus.

An obvious objection to this claim would no doubt be that descriptors are also supposed to denote concepts. However, Manuel Zacklad rightly underlines how much these concepts are in fact the direct expression of vernacular languages, a feat that is reflected through their intrinsic variability. Because it is difficult to isolate an autonomous conceptual level from its linguistic representation (or rather, from *any* linguistic *expression*), a shadow is thus cast on the real scope of the referential semantics traditionally ascribed to descriptors.

One way to slightly alter this judgement would be to stress the difficulty to choose between several terms in order to represent a concept.

Cf. Zacklad:

> "the creators of a thesaurus will adopt a linguistic expression, the descriptor, and consider it, *ceteris paribus*, as the best surrogate of the intended concept" (Zacklad, 2007, p. 6)

However, the exhibited similarities, relying on a semantic apparatus that includes synonymy and translation (intra-linguistic as well as inter linguistic), they finally, correspond to the much criticized notion of "meaning postulates", the one advocated by Rudolf Carnap in his book *Meaning and Necessity,* and famously criticized later on by his student, Willard Quine. (We lack space to further elaborate on the locus where inferential semantics, meaning postulates, and the principle of literary warrant in library sciences all collide).

As regards tags, whereas relations are *de facto* established between them, contrary to descriptors, they're never specified.

Of course, this does not take into account systems which tap into thesauri to add more structure to a folksonomy. It could nonetheless be argued that label idiosyncrasies escape all attempt of regiment. Moreover, for the same reasons but from a referential point of view this time, the denotational status of label is an intractable question (cf. *infra* §3.4 sq.).

3.2. Subject Headings

A subject heading is no simple word. It is a compound lexical entity singled out from a documentary language – instead of a natural language. In other words, a syntagm created after the coordination of several descriptors. The distance between natural and artificial (documentary) languages is to be measured, on a semantic level, by the distance between the models (the model-theoretic sense of the word) used to clarify the meaning of the lexical items discussed. As Helen Svevonius (2000, chapter 8) noted, "subject language terms differ "referentially" from words used in ordinary language. The former do not refer to *objects in the real world* or *concepts in a mentalistic world* but to *subjects.* As a name of a subject, the term "Butterflies" refers not to actual butterflies but rather to the set of all indexed documents about butterflies" [my emphasis]. What is at stake in formal approaches to semantics is, amongst other things, what Emon Bach (1989) called the "natural language metaphysics", a critical analysis of how linguistic components denote (to which entities they refer to). As regards subject headings, the putative domain of their model (the set of all the things they denote) is composed of documents about the same content; which share a common theme.

The norm AFNOR NF Z 44-070 adds this:

> Each subject heading corresponds to only one subject, simple or complex. One document can have more than one subject thus prompting the wording of a plurality of subject headings. (AFNOR, 1986)

Beyond the well-known challenge raised by lexical entities to formal semantics, every attempt to interpret a language according to only one model is doomed to shed light on the design process behind the production of that language rather than on its effective use. Tags, on the other hand, are free from this limitation. One need only point at *"to-do"* tags (Bach, 1999, Marlow et. al. 2006), traces of actions related in one way or another to resources without the need for them to denote them, to dismiss any attempt to equate subject headings and tags (labels).

3.3. Keywords

Admittedly, it is difficult if not altogether impossible to reach a consensual agreement regarding the definition of keywords. The very success of this notion comes with a high price to pay: unrestrained fuzziness. From "natural language" queries typical of search engines to keywords as defined in library sciences the gap seems almost preposterous. That is why only this well-established usage will be our concern for the time being. As the norm AFNOR NF Z 47-102 puts it, a keyword is a "word or group of words picked up either from the title and the main body of a document or from a document request, in order to describe its content". A keyword, albeit bound by its origins to natural languages, is the result of a careful analysis of a written document. Tags, thanks to the versatility of the digital environment of the WWW, best characterized by the word "hypermedia", are used with a variety of resources that far exceeds written documents from which keywords are extracted. For this simple reason, the label of a tag cannot be assimilated to a keyword. Huge as it is, the variety of written documents is nothing compared to that of tagged resources since the latter may indeed include photos, audio recordings or even real-world events as we shall see. Let's go even further by noting that the impossibility to extract keywords from audio recordings (computerized metadata being another thing) or world events (the website "Upcoming.org", for example, allows one to [machine] tag events) stands in sharp contrast with the main feature of tagging: the opportunity to add strings of characters to a resource, a resource being, according to the W3C "whatever might be identified by a URI" (RFC 3986). A logic of expression, wherein content is *added* rather that *parsed*, takes over the one behind the definition of keywords. With the emphasis shifting solely from description towards expression, we also get the means to distinguish "real tags" from fake ones. Those applications which quite spuriously pretend to produce *tags* from text analysis are in fact only able to *extract* keywords whereas tagging is, in a very large measure, tantamount to *adding* absent or rather *extrinsic* content to a resource.

3.4. Labels as Blank Spaces

The most striking contrast with previous existing bodies of practices and norms, from the point of view of professional indexers, must have been the shift from *a priori*, controlled indexing to the current freer forms. An immediate consequence of this newly acquired freedom, in a nutshell, is that labels are no longer terms of a thesaurus, subject headings, or even words, but all this at the same time and even much more than that, including, among other things: triple and/or machine tags, URLs, smileys (iconic representations in general), messages written in Windings, code, etc.

Nothing prevents infusing a sign like "@" with an idiosyncratic meaning, to the extent that one fears not dwelling at the frontier of the idiolect and distance oneself with the received sociolect. Machine tags are another borderline case where machine-readable computer code, designed specifically in accordance with blue-ribbon Webservices, supplies human-understandable information owing to an artificial syntax that mimics that of natural languages (those technological devices we qualified as "artefacts of meaning" are deeply teratological in nature from a linguistic point of view; they deeply alter our language games in a way previously inconceivable).

If the freedom to choose one's own labels that is now offered to users was to be properly acknowledged then there would remain only one conclusion to be drawn. Contrary to subject headings or descriptors whose semantics is rigidly established in line with a single model of interpretation or through a well-ordered lexicon – thanks to a small number of relations established and postulated in order to evacuate every remainder of ambiguity – labels written on tags are able to comprise various entities, linguistic or not, thus forbidding all global theorizing on the semantics behind their use. In other words, *the label of a tag is a blank space that is fit to accommodate any sort of inscribable entities* (as far as the system is made to be compliant.*)*. As such, it is therefore completely devoid of any fixed semantics (such a question doesn't even have a sense as regards tags understood exclusively as technological devices).

4. Resource Indeterminacy

Rashmi Sinha (2005), in her widely quoted analysis of the cognitive dimensions of tagging, undertakes an explanation revolving around the idea that it constitutes a softening of categorizing, one whereby directories and their well-publicized shortcomings are no longer needed. According to Sinha, in order to make use of directories one has to pay a tremendous cognitive cost, an outlay that may be easily dispensed with thanks to the low-entry barrier of tagging. Of course, this is not to say that everyone by simply using tags is automatically rewarded with the skills of a professional indexer. Yet, indexing as an activity, now that it allows a higher rate of errors, redundancies, nonce words and hapaxes, is certainly made to suit, more so than any preceding comparable system before, the layman's needs.

Such a vision, though appealing, is nevertheless missing a fundamental element. In spite of its seeming simplicity, tagging hides a repeatedly overlooked difficulty. It has been painstakingly argued that tags instantiate relationships of reference and access (and many others but we've been only focusing on these two). Now it's time to ask another question: to what? Real entities? Digital resources? Far from obvious, the answers to this question echo in a large measure those raised by dereferenceable URIs on an information network such as the WWW. The various forms of "aboutness" displayed by tags suggest the possibility that each of them corresponds to a different kind of entity.

As regards those available on the Web, they qualify for a large part as "intentional artefacts" – to borrow a concept from Daniel Dennett (1990). The way we deal with such artefacts is unavoidably tainted by their aboutness. An image might be *about* something as well as a text or an audio recording. Hence, whenever a user is tagging an image for instance, he is at the same time asserting something about (and referring to) an entity located either on the Web and/or outside of the Web (the image *per se* vs. what it stands for), and leveraging the possibility to *access* it afresh at any time (the picture located on a given website where it has been allocated an identifier which is also – potentially – an address).

This relation of access itself is just as indeterminate as any linguistic description.

(In order to circumvent the phenomenon called *deixis*, the concept of «*deferred ostension*» is sometimes used to distinguish the index from the referent, *pointing at* from *referring to*. It follows from this perspective that "*monstration*" (to use a French word) is located halfway between technological access and linguistic reference: an area where language embodies gestures – and extends as far as to incorporate tools and artefacts.)

True, the resolution of a URI into a URL is the result of a controlled negotiation between agents dictated by the technical standards defined in protocols (most notably HTTP). Once a page (what exactly is a Webpage remains a difficult question.) has been generated though, the exact nature of what has been described thanks to the label of a tag remains open: is it the page itself, or a part of it? And which one? A blog post, a commentary or both? The hyperlink itself? An action that should be prompted by accessing something like a resource?

We've already alluded to the intangible nature of resources. Strictly speaking, to be a resource is tantamount to being just about anything. Still, since only informational entities are available on the Web, a notion was eventually forged in order to single out a subclass of resources in a more explicit fashion; that of "information resources" (Halpin and Presutti, 2009, 521-534). Information resources are described as resources which can be represented online (an information resource might be "the weather today", its representation would then, for example, be a webpage generated at a given moment containing the daily weather report). The role of agents is, for each HTTP (*Hypertext Transfer Protocol*) request, to generate a representation in accordance with a set of technical parameters. Of course, agents have their say in determining the kind of representations the user of a tag is given access to. However, from her point of view, what the tag is able to retrieve is inescapably intermeshed with what the label of a tag refers to, except when the two strongly differ (a real-time event can hardly be taken for a webpage describing that event, while a given document can, *ceteris paribus*, undergo conversion from PDF to HTML or any other format that maintains its legibility without too much alteration).

Hence, whenever access and reference are jointly associated, as is the case with tags, what *the user* may have access to (whenever tags play a pivotal role in her browsing) happens to be filtered by the labels she used when she referred to it (and in some cases it happens that access is given to just what a label refers to, for instance when the representation of a resource is indistinguishable from the resource itself; though a difference remains in principle which explains why reference may be made specifically to a resource and not to its representation, thus inducing more confusion – or complexity, depending on one's standpoint).

5. Conclusion

We postpone the examination of this question for further research, in the hope that a full blown inquiry into the hows and whys of this (at least) "bi-dimensionalist" (to distinguish it from two-dimensionalist approaches to meaning) approach to meaning, rooted in language, semiotics and technology, would help to identify and evaluate enough relevant case-studies for a theory primer to emerge. Comparable phenomena of indeterminacy have been already identified in natural language use under the name of "systematic polysemy". Yet, it remains to be seen whether or not lessons learnt from studying this issue are relevant – let alone transposable – when it comes to such artefacts of meaning as tagging systems. If technology is truly altering the rules of our languages games the task that is awaiting us is to honestly assess to what extent.

SUBSTANCE-ENERGY-INFORMATION MODEL IN THE ANALYSIS OF COMPLEX SYSTEMS

GEORGE CEAUSU

The *SEI* model (Substance-Energy-Information) is made up of three invoked traditional notions in good Aristotelian filiations, in the capacity of *categories*. As they are notions with maximal extension, indefinable by other proximate kinds, the categories have a different statute compared to other notions. To the one encountering in his domain specific real notions, such as that of "chemical substance", it remains still unclear whether the *substance*, generally, is a real or fictional notion. On the other hand, the socio-humane sciences practitioner, working mainly with texts and behaviours, often despises the model of a "reality" derived out of categories, preferring constructive schemes with parameters of communication and cognition (emitter-receiver, significant-signified, direct or figurative sense) – schemes wherein the model terms are not retrieved.

1. Systems of Real or Sermocinal Sciences

In the capacity of the traditional philosophic category, "sowing" the scientific meanings too, the substance (*substantia*) is what it is underneath – *sub-stantia* – at the bottom of changing properties and of modifications; permanent carrier of objectual and, at the same time, essence. The Latin transcription of *ipokeimenon, ipostasia, ousia* feeds the comparative debate between the Greek philosophy and the Latin one. There are prime substances (individual work) and secondary ones, the first having a privileged role – in the sense that the secondary substances are logical or physical combinations of the prime ones. The four aggregation states – solid, liquid, gaseous, plasmatic – also constitute a systematizing base of the physical effects depending of the nature of the object subjected to the influence of effect entry (Sandu, 1994).

In the case that substance is passive, a support, a memory of the events, then the dynamics of the respective even is disputed between information (*an intelligent active*) and energy (*a non-intelligent active, stihinic*). This perspective is revaluated by artificial intelligence – within which the human, animal and computer simulated intelligence are treated in a common way. Coming out from the Hellenic *energéia* (activity), the energy takes over the movement category, of material and spiritual (inter)activity. Within the physical systems, the energy is their capacity to carry out mechanical work when it undergoes a transformation from one state to another. It is kinetic (traceable in movement) and potential (transformable in movement); aspects closely related between them by an additive relation in the law of energy conservation and transformation. Other terms from "etymological nest" of *energy* could be force, power, impulse, acceleration, movement, rest etc. Regarding the energetic field, we should have an exigency: determining the *wave equation* – or at least, empirically, setting some geometrical characteristics of its wave front in a Euclidian, non-Euclidian, or fractal geometry.

Regarding the informational section, we should work either with *signs* (informational realities that admit the semiosis, e.g. evidencing the psycho-linguistic and socio-linguistic aspects, mentioned in the previous section), or with symptoms (set of signs over which logical deductions can be made), or with signals (defined functions over sets of indistinct data). Information has its elucidated structural aspects (information theory, cosmic-sciences) and non-structural aspects (semantic) within the representation theories' concern. Putting into form, information is subsequent in the historical eon of the form category – *eidos, morphé* – that, in a good tradition, represents the matter correlate. Existing in act or in potency, the forms determine even the existence of things regarding the aspect of their species. The information, a newer concept in knowledge order, with categorical value, knows real refreshment, beginning with the 1930s (founding the deductive systems and the calculability theory) and with the construction, 15 years later, of the first computers.

To the three concepts of the *SEI* model title, there are added, according to Paul Constantinescu (1990), two coupling ones: *i) matter*, *SE* coupling resultant; *ii) the field*, coming out from *EE, II* or *EI* couplings – the last is the coupling of the energy-information or information-energy type. The field designates, *in fact,* knowledge of labile limits between energy and information, plurality of energies, and information systems. Since Aristotle's *form* of *Fizica* referred to the whole (by form-cause and form-principle), information has been a chameleonic and pluralist concept under a formal aspect, which explains its modelling success by sets of fractals or sets of “strange numbers”. The derived notions of the model are systematized in *Table 1* where, the rows show the first element and the columns show the second element of the coupling.

Table 1: Derived notions of SEI model

	S	E	I
S	“secondary substance” (compounds)	“ondulatory matter”	“process output data”
E	“corpuscular matter”	“compound energy”	“field”
I	“process input data”	“field”	“derived information”

The informational system is an ensemble of human and capital resources invested in view of collecting and processing the data necessary for information production – data to be used on all levels of control and decision (Oprea, 2005, 35). The reference from where we extracted the definition refers the economical systems, but also the ensemble of texts to be proposed is animated by the request for a more accurate “situational description” – until the making up of a knowledge basis with a relative homogenous structure. Based on the analysis of some similar systems, a new system in the field can be designed (first a logical one and then a physical one), creating, in turn, formats (forms) and reports; user interface; programs and procedures (Oprea, 2006, 15-18). Starting from a data and knowledge set, we designed *an informational system* containing idea streams and basic concepts in philosophy (Ioan, 2007).

The composition of the *informational systems* is the following:

$$SI = Hw \cup Sw \cup Fw \cup Lw \cup Bw,$$

where we applied the notations: *Hw* = *hardware*, *Sw* = *software*, *Fw* = *firmware*, *Lw* = *lifeware*, *Bw* = *brainware.*

The informatic system is that part of the informational system that disposes of the *automatic* procedures (algorithmic) meaning that we can appeal to an identical scheme for the informatic system too. By comparison with the automatic procedures, the *manual* ones contain instructions for the engaged personnel in the system. It is accepted that the distinction between the informational system and the informatic one, as a system (mostly in the socio-human sciences) cannot be entirely programmed. We rather should say that it looks like a Swiss cheese where the solid part is composed of automatic programs and procedures and the holes of manual procedures.

The analysis of informational systems implies, firstly, delimiting the vocabulary and the "regulations" related to the language, precisely an inventory of units (phonemes, morphemes, lexemes etc), a relation structure provided with an amount of application rules and a corpus of texts (Lobiuc & Irinescu, 2002, 55). But, in the theory of databases, we need a *vocabulary-text-discourse* triad. The language is a set of signs mainly orientated towards the communication process. In fact, the main functions of a language are *communication* and *cognition*. A language is based on *a vocabulary* where the means of all the terms are clarified. In order to communicate with others, we need a text, e.g. a coded sequence inside this vocabulary generated by correct grammatical rules. Such a conception of the difference between vocabulary and text also serves us in working with signs other than words and *phrases* (word sequences). Because there are also languages of a pictorial, musical, gestic etc. type, inside which, starting from signs with a certain meaning, we can compose corresponding *textures. The discourse* is the improved text with the most various social behaviours. The *vocabulary-text-discourse* triad should allow further making of order in the multitude of program systems but also in texts. There are more language types on current computers: verbal (articulated sequences: words and phrases), imagistic, filmic, sonorous, manual (of control through the mouse) etc.

Algorithm and *data* ideas are primitive in informatics providing descriptions, not definitions. On the other hand, the primitive notions of mathematics (at least in the so called "ensemblist" variant) are the set and the element, and the primitive relation is the relation of affiliation "$\in$". These notions are fully "evaluated" from an informatic point of view too. *The algorithm* is a sequence of operations (instructions and commands), orientated towards solving some problem, written in a preset language and basically having a cybernetic model. *The data* is a standardized representation that can be subjected to an automatic processing. In the theory of effective calculability, the algorithms have a functional role, and the data have a structural character; from another point of view, the algorithms represent the dynamic component, and the data the statistic component of a program system. As with the algorithm concept, the *program* one is very broad in informatics, often being treated as being primary. The experience imposed a formula of the *program* (as a component idea of the computer, together with *equipment*): *program = algorithms + data* (Wirth, 1976).

In our approaches orientated equally to life (real) sciences and to socio-human sciences, we preferred to work also with two counterbalances of the *real* term, namely *the fictional* and *the imaginary*. The real and the fictional are in a relation of contrariety (there are no things real and fictional at the same time), and the real and the imaginary, in a relation of sub-contrariety (there no things real or imaginary at the same time). We notice that the real and the fictional behave also like universal syllogistic judgments, accepting as referential all the other combinations between truth and false (except, the *truth / truth*! combination). Likewise, the real and the imaginary behave also like the particular syllogistic judgments, accepting as referential all the other combinations except the *false / false*. We also may introduce the distinction between real and fictive (unreal), building up a contradiction relation between the two notions. Likewise, the negations of the first two terms (*non-fictional* and *non-imagistic*) help to determine the circumstances of the real.

Among the *correlative* terms of the real (not opposed!), the most interesting one seems to be the *duplicate of the real* or of *twin-world* – like in the mental experiment with the "Twin-earth" of Hilary Putnam. Hilary Putnam gives the example of the famous duplicate of the "Twin Earth". Let T_{water} be Terra before the discovery of water structure and T_{H2O} be the same planet after the discovery the H_2O formula. There are considerations of physical or sociological order supporting the relation $T_{water} = T_{H2O}$, but also considerations can be found out by which $T_{water} \neq T_{H2O}$, and in this way we may delimit two different worlds, not only one; moreover, nobody stops us to construct the P_{water} or P_{H2O} notions, where P is an planet other than Terra, possibly a fictive one. It is a procedure by which the W world can be improved with more duplicates for which we accept either the copy on scale (homology) or the selective copy on the set of feature structures (analogy). Admitting three such types of duplicates, namely sociologic, physical and psychic (W_s, W_f, W_p), we reach a fantastic typology with 12 system inputs.

There are also two interesting correlative terms, the one of *extended real (virtual)* and the one of *constrained real*, denoting enlargement efforts, respectively narrowing the real. The term *artificial* seems to be paradoxal for us, being in *tautological* relation to the *real*!

A basic supposition of the literary discourse, the "text as semiotic invention", encourages the emergence of some opposites of the real or fictional, imaginary and unreal type; increasing the access difficulty to a "real" conceived as a picture of the contemporary world to the knowing subject. In the first part of the work, considering the distinction between real sciences (realist) and the sermocinal ones ("of the discourse") at the level of the revolutionary notions of *system*, evidencing the fact that the semiotic parameters usually "camouflage" the *SEI* model coordinates in the behaviour of an intelligent agent who is often called "human being".

2. The Systems in Socio-Human Sciences

The system complexity is in a relation of direct proportionality with *information* accumulated by it. There are complexity degrees directly correlated with the level of description and understanding of the highly organized systems. Another definition uses the term of *undeterminable* as synonym to complexity, asserting that the systems in which the algorithm is approximately as big as the structure itself are undeterminable (Chaitin, 1964). Differently, Kurt Gödel saw complexity as being synonymous with *undecidability* and his principle says that

> if it is easy to speak, within the language, about the language itself, it not so easy to understand how a certain phrase on numbers can refer to itself (Hofstaedter,1977,54)

because no mathematical proposition set does contain in itself the possibility of self-checking.

For Werner Heisenberg, on the contrary, the complexity is similar to *quantum indetermination* according to which the place and the impulse of an elementary particle cannot be measured at the same time. It recalls an idea known under the name of *complementarity principle* in order to solve this dilemma that human senses have to cope with. Intuited by Ludwig Bolzman and theorized by Niels Bohr, the principle asserts that, in order to apply various aspects of a given phenomenon, contradictory hypotheses can be used. For example, between the description of electrons as particles and, respectively as ondulatory phenomena, there is the same relation. The two descriptions are catalogued by Niels Bohr as *complementary*. This term firstly shows that the descriptions are reciprocally excluded and consequently cannot be concomitantly realized; secondly, it shows that they reciprocally completed, only together representing a coherent description of the experimental known effects.

Delimiting the survey object of a science (out of the existing tens of thousands) becomes in this way necessary. Not accidentally, in UNESCO classification the problem of delimitation between main and secondary domains arises, because the ontological freshness of it escapes any delimitation, any placing in drawers. A table with the main scientific disciplines (in which appear also domains restricted by research) can be found in a reference on history of science (Neuman & Nicolau and Schor, 1983, 590-599).

In a current classification of UNESCO, over one thousand main sciences are operating! In this respect, delimiting *the physical object* creates big problems to a *knowing subject* – agent sank in an *environment* where the knowledge *objects* are spread. The knowing subject surpasses the individual knowing status, entering in correlation with other subjects. In the 20th century, because of the perspective induced by ondulatory mechanics and by the elementary particles theory, the subject / object delimitation suffers: quite many experiments illustrate how coarse our senses are and even the devices designed for their extension.

Our relation with the (elementary) particle and with the wave points out a realist behaviour in the world – "the world" being a total concept that requires being reconstructed by comparison, often irrelevant to its parts. When we say *spatial-temporal continuum*, we have already defined an independent and minimal system of categories to figure the sophisticated notion of "physical reality". At this point, correlation of *SEI* system with the spatial–temporal continuum is compulsory.

In the previous section, we detected in fantastic and SF literature, three types of "twin" worlds (relating to *our world*): physical, psychic or sociological. The "twin worlds" reaffirm the identity principle of Leibniz by relating to the *SEI* model. The largest relation conceived in the spirit of the Identity Principle is:

$$S \equiv E \equiv I$$

(where "≡" denotes a weaker relation than identity, e.g. equiordination of notions) in order to describe what, under energetic-informational relation, we call "physical reality". We shall come back on the above relation but we should set a system of the "Occam's razor" type (of independent and consistent notions) in order to describe reality. Further on, we shall admit the thesis:

STC = Space, time and continuity are sufficient in order to describe physical reality.

When we say *spatial-temporal* continuum we have already defined a system of the "Occam's razor" type in order to figure the sophisticated notion of "physical reality". We do not think as necessary to argue why time and space can not be omitted in a system of reality description.

We shall remind in this paper a short description of inter-disciplinary as principle transfer between domains and of trans-disciplinary as concept transfer (Ceauşu, 2007); a description that can be modelled by a pseudo-Boolean logic of concepts. We distance ourselves in this way from the approach Basarab Nicolescu gives to inter-disciplinary as transfer of method, but we approve his or other experts' diagnosis related to the increasingly profound scission between real sciences and socio-human ones. The approach of inter-disciplinary implies both the scientific versant and the religious one; the distinctions between the object and subject of knowledge emerge also by the disjunction between science and religion.

Further on, by *organism* we should designate *the complex system made up of body and psyche. The organism* is a biological or medical notion, not always accepted by itself in the common sense. Some are monists, other dualists and other trialecticians. We can work also with "wider" models (numerically) of the organism but we stop here. There are three concepts (organism, body and psychic) confronting the identity principle. Even the three basic notions of the *SEI* model confront willy-nilly with the principle of identity.

There are many identity theories in mathematics and logic that mean the following: in the theory language, there is a predictive symbol of 2nd order, let's say *P*, named *equality*. We shall agree on writing $t = u$ instead of $P(t, u)$.

Of course, among the axioms of a theory with equality should exist some of them implying the existence of theorems:

1. Predicate *P* holds the properties of an equivalence relation (reflexivity, symmetry, transitivity);

2. $\forall u_1 \ldots \forall u_n \forall v_1 \ldots \forall v_n (u_1 = v_1 \wedge \ldots \wedge u_n = v_n) \supset f(u_1, \ldots, u_n) = f(v_1, \ldots, v_n))$, for any functional symbol f of the order $n > 0$; (3) $\forall u_1 \ldots \forall u_n \forall v_1 \ldots \forall v_n (u_1 = v_1 \wedge \ldots \wedge u_n = v_n) \Rightarrow R(u_1, \ldots, u_n) = R(v_1, \ldots, v_n))$ for any predicative symbol *R* of order $n > 0$.

The comment of Leibniz regarding the *identity principle* (if $a = b$, then $P(a) = P(b)$ for any predicate *P*) remains open and coincides with t_R property (considered for a single variable).

The reverse relation, namely:

$a = b$, *as* $P(a) = P(b)$

is named by Leibniz "indiscernible principle", not always being valid for a function.

It results that equality, far from being a common equivalence, complies with some criteria making it able to "attack" formally the material identity. The two principles of identity and of indiscernibles show the fact that theories with equality are at least of 1st grade (with quantors at individual variables), but, most frequently, of the 2nd grade (with quantors at sets). So, the calculus with predicates would by a *minimal* model for a "finer" representation of reality.

In a Boolean calculation with notions, "S ≡ E" identity extends over a world made up only of substance and energy, so we could not have but "substance ≡ substance" and "energy ≡ energy"; not accidentally, in the syllogistic calculation it is also introduced that the existential supposition on notions, adding to the above described world, in black and white, and "grey" nuance (marked in Venn diagrams with asterisk, so that it would be of the sort either "substance" or "energy"). The existential supposition hangs also over notions ("there is energy") and on relations between them (there is not "non-created energy" in science).

The considerations of this kind announce a larger discussion of mind philosophy trends around a theory with equality in which we could accept that:

"S ≡ E", if S and E describe the same aspect of reality,

where *S* = *"substance"* and *E* = *"energy"*. The interpretation under the intentional aspect (and not extensional) of the *"body" ≡ "psychic"* relation should generate a discourse on the way in which a theory of notions can be founded based on axiomatic theories of sets, such as the Zermelo-Fraenkel theory. The naïve theory of sets does not suit, as it does not suit from one point further, any "semantic of possible worlds" (in the sense prefigured by Clarence Lewis and elaborated by Saul Kripke), because it conflicts with the Bertrand Russell paradox (of the set of self-contained sets). Here is not the place to make a case to develop this idea.

In the previous section, we detected in the fantastic and SF literature three types of "twin" worlds (by relating to *our world*): physical, psychic or sociological. The "twin worlds" reintroduce the identity principle of Leibniz by relating to the *SEI* model. The largest relation conceived in the spirit of the Identity Principle is:

$S \equiv E \equiv I$

(where "≡" denotes a weaker relation than identity, e.g. equiordination of notions) and the derived relations are

$$S \equiv E; E \equiv I; S \equiv I$$

when the question of coexistence arises between two terms (e.g. the *substance* and *energy* describe the same aspect of reality, for example).

Out of the wide inventory of psychological methods, a part of them in the lead with "introspection methods" come to be contested. The social psychology brings also on the stage anamnesis, enquiry and questionnaire methods, unspecific to the real sciences. The sociology also brings into the horizon of scientific methods the enquiry and the opinion poll (Rotariu & Iluț, 1997, 44-70). That is the reason why the introspection method, giving to the psychology the aura (maybe not deserved) of "objective science", could be contested and the psychology, as a "cultural encyclopaedia", could be sunk into philosophy*.

* Translated by Decebal Necula

PRINCIPLES OF THE BRAIN-COMPUTER ANALOGY: A BIOPHOTONIC APPROACH

TRAIAN-DINOREL STANCIULESCU

1. Premises

The complex process of cognition could be indirectly defined using its three functions, correlating the human being's qualities of *significans, cogitans-loquens* and *sapiens-sapiens,* namely:

- *the symbolic-semiotic function*, concerning the "essence discovery" of objects and phenomena, in terms of essential cerebral operations (such as representation is);
- *the cognitive-linguistic function*, supposing the assumption of the essence of reality in terms of a certain language (nonverbal or verbal);
- *the logical-discourse function,* concerning the coherent mental reproduction / modelling of the reality.

Each of these three competences corresponds to a progressive level of human cognitive-communicative performance. In the present context, we will take into consideration only the first function mentioned: the symbolic-semiotic one, as an essential step towards natural and artificial intelligence. This function emphasizes the human being's capacity to mentally represent the exterior or his / her interior reality, by using some substitutes of reality (symbols / signs) and to realize operations with these substitutes. Representation is essentially connected to memory, to the brain's physical / bio-chemical support / signifier (as a "cerebral hard") respectively, on the one hand, to the activation of this support by using a specific program / signified (a "cerebral soft"), on the other hand. That is why we could say that to explain the mechanisms of memory and representation is in principle equal to explaining the main mechanisms of the cognition process.

Because the conventional scientific theories were not able to entirely clarify these problems, we will use for this purpose the unconventional hypotheses of a new emergent science – *biophotonics*, developed by professor Fritz-Albert Popp and his team (1989) – and assumed by us as the study of the processes of generating, storing, releasing and interacting emissions of biophotons ("ultraweak bioluminescence") at the level of all the leaving systems, in their quality of "biological lasers".

2. "Biological Lasers" Theory: Main Hypotheses

The central postulate of the "biological lasers" theory (BLT) – developed by us as an original contribution to the foundation of biophotonics (Stanciulescu & Manu, 2001) – is that, among other functions, biological structures also accomplish a function ignored by the conventional sciences: the optical function, describing the photo-sensibility of the living structures from the non-linear optic perspective.

1. *Structurally*, from this point of view, at the level of the human body, the "metamorphoses of light" – which pass from the physical state (photonic) to the biological one (biophotonic) – are ensured by the following two systems:

- the molecular semiconductor-type system, represented by the complex phosphate / linked water / molecular oxygen at the cellular membrane level, which generates the bioluminescence phenomenon, that is the laser-type emission of light;
- the organic structures with liquid crystal properties, such as the cholesterol molecules and phospholipid from the cellular membranes, which determine the physical properties of the laser type emission (coherence, monochromaticity, intensity, directionality of the incident light) through birefringence, polarization and selective reflection.

2. *Functionally*, the two premises are the basis of the BLT's major hypotheses:

a. The correlated action of the molecular semiconductor "biological-laser" and of the organic structures with liquid crystal properties allow us to argue the idea that the molecules, membranes, cell nucleus, cells, organs and the whole organism function just like some "biological-laser" systems, both "linked" and "intricate". On the one hand, all these systems present evident structural analogies with the technical lasers: optic resonator, active substance, optical pumping source of light, cooling system, electromagnetic director field, etc. On the other hand, the functionality of these biological complexes generates a bioluminescent phenomenon, having all the mentioned properties of the "natural laser" emission.

b. The structural-functional particularities of the six types of "biological- lasers" existing in the living complex body enable us to emphasize the correlated effects of four types of phenomena: Biochemical, Electric, Magnetic, Photonic (BEMPh) effects. These phenomena are characteristic of all the biological processes of the living body.
c. By the energetic accumulation permitted by the presence of light, by the energetic discharges which take place in absence of light respectively, it is possible to define two essential cycles which maintain the vital processes on a dynamic equilibrium: "daily cycle" (of light) and "nightly cycle" (of darkness).

*

Such processes and mechanisms permit us to formulate some unconventional explanations concerning: the nature / structure of the DNA helicoid and dextrogir form; the stocks of biophotons in the micro- and macro- biological "traps of energy-information" (in DNA, mitochondria, etc.), and the ways of their release, the redefinition of the health / illness state as a function of transparency against light, etc. The common pattern of all these processes is the "living light" (bioluminescence).

By metamorphoses of the "natural light" (visible) into "biological radiation" (invisible), deviated to ultraviolet, a bioluminescence emission is generated at the level of each biological structure. This "biological-laser" energy is defined by: the properties specific to the laser light; decreased impulse frequency of the monochromatic radiation and increased energy; decreased speed caused by the successive penetrations of light through the dense mediums of the biological organism, etc.

2.1. Neuro-Semiotic Principles of the Brain-Computer Analogy

The correlation between biophotonics and the structural characteristics of the human personality permits us to draw an original explanation of some still unknown or less known aspects concerning neuro-semiosis and logo-genesis.

These aspects justify the interest in a "semiotics of light", because:

- *ontologically*, the double nature of light could be associated with the dual structure of any sign, the light / sign having the coherence / resonance relationship as a mediator;
- *epistemologically*, light constitutes the object of the same archetypal knowledge, for which semiotics is an organ.

It follows that between the light generating cosmos and the human receptor a "significant resonance" is instituted. This resonance involves the bio-psychological and social human components philo-genetically and onto-genetically. The present paper studies them and formulates some still unconventional hypotheses of an integrative "semiotics of light" (Stanciulescu, 2003).

Synthetically, the above mentioned ideas permit us to integrate a still ignored function of the biological structures – *the optical function* namely – into the context of semiotic explanations. Therefore:

a. On the one hand, we could find in this option the conviction that "a biologically based semiotics will allow us to get a glimpse into how the body interacts with the mind and culture to produce signs, messages, thoughts, and, ultimately, human behaviour" (Danesi, 1998).

b. On the other hand, we could answer – from the particular biophotonics point of view – to Thomas A. Sebeok's (1972) question on "life and human essence", by stating that human life is the synergy of some specific biological, psycho-logical and social features.

*

From the perspective of biophotonics and psycho-neurology, it is possible to develop in the present context a semiotic model on the brain-computer analogy (hardware → signifier, software → signified). This model is able to explain the main hypothetical principles which govern neuro-semantic activity at the level of the three bio-cybernetics brain blocks, which computers also have, respectively the block of: (1) *information reception-codification*, (2) *information processing*; (3) *command and control*. In the present context, we will enunciate only some of these unconventional principles, belonging to the three "cybernetic blocks", which could originally explain the brain's performances and, at the same time, its possible analogy with a super-computer:

1. *The principle of energy-information unity*: any impulse received by the (nervous) cell will be homeomorphically transformed into electrons and biophotons (bioluminescence), due to the piezoelectric properties of the membranous liquid crystals. This mechanism suggests that the significant connection between the referential energetic-information properties (mechanical, visual, caloric, etc.) and its correspondent cerebral sign (as representation, information / signified) is always mediated by one and the same electric and electromagnetic energy (as support / signifier).

On the one hand, computers use the electricity as a source of energy, which could be obtained by different kinds of energetic transformation (e.g. by photoelectric effect), if the computer is dotted with such an installation. Afterwards, the electricity is transformed into visual and sound information in the computer's hardware. On the other hand, the internet already uses the optical fibres for transmitting the information by photonic signal; it is exactly what the brain does, using its nervous net as some optical fibres, for transmitting biophotonics signals (information) from the analyzers to the cerebral and from here to the effector organs. Consequently, from the perspective of biophotonics, all the nervous net is organized as a very complex internet system.

2. *The principle of redundant propagation of information*: the same received information is transmitted "step by step" to the brain, by the complementary ways of BEMPh phenomena: by biochemical transmitters at the synaptic level, by membranous electric potentials and by the magnetic cellular fields, by the biophotonic fluxes (which penetrate the interior of the cells, due to the membranous liquid crystals transparency). All these ways transfer the specific information of the referential (signified) to the brain support (signifier): chemical (liquid crystals, neuro-transmitters, etc.), bio-physical (electric and magnetic field, photonic flux).

In the same way, using the same "hardware", the computer is able to generate the same effects by using different "software" registered in its memory or by using different commands in the frame of the same program.

3. *The principle of information simplification*: the complex of exterior information is decomposed in elementary information (such as frequency, intensity, phase, form, distance, pressure, etc.) at the level of the different sections of the neuro-analyzers and are received / registered in such form at the level of the posterior cortex. Consequently, each complex of stimuli belonging to a certain referential is first represented as an assembly of simple components (primary signs) at the cortex's level, and after that as an integrative sign (having visual, sonorous, tactile properties etc.).

Similarly, all the complex functions of the computers are realized by using the informational "points of information" – the bits – and its octets, the bytes.

4. *The principle of cerebral hierarchical functionality*: due to the holographic organization of the brain, biophotonics could explain why the major functions of the brain (such as memory, for example) are both "localized" (at the level of cortex, where the simple information is received) and "not localized", because of the distribution of the memorized information / signified at the level of the entire cerebral mass,

having as signifier a substantial (the membranous liquid crystals) or energetic (bio-electromagnetic waves) support.

In the same way, due to the "hierarchic" construction of the computer, if some components of its memory are affected, they can be substituted by others, allowing the action of certain functions. Of course, such a substitution is possible only in certain limits.

5. *The principle of the holographic reception and information processing*: this mechanism is specifically activated in two circumstances: (a) at the level of the peripherical analyzers, where the signals / signs are registered as some "holographical points-object" (optic, acoustic, etc.), as some iconic "reproductions of the referential; (b) at the level of the central nervous systems: the holograms are realized in the hole mass of the brain, being generated at the different profound levels of the brain, by the interference of the different (simple) information (coherent bio-electromagnetic waves) which describe the referential.

This principle is not directly involved in the computer's functionality. But, the multilevel and redundant structure of the computer, able to supply a missing function by another one, could be considered as being a kind of "holographic function".

6. *The principle of the holographic memory's functionality*: the membranous liquid crystals could be considered as the substantial support (signifier) for the information (signified) carried by the stimuli (referential). Due to their diamagnetic properties, activated by the incident light (as a "sign" of the incident stimuli), the membranous liquid crystals are able to change specifically their position, configuration etc., and to keep / memorize in this way (morphologically), for a certain time, a specific form (to inform = to put in form).

A computer's memory also uses the properties of the magnetic support (of the hard-disk, diskettes), and the properties of the micro-chips (based on the properties of the silicon micro-crystals).

7. *The principle of the holographic representation*, by two categories of stimuli: (a) (in)direct excitation of the brain, by a *physical signal* (electromagnetic, for example), able to generate at the level of the signifier some resonant phenomena (energetic, predominantly); (b) *the stimulation by sign* (word, gesture etc., e.g.), able to involve a certain resonance (informational, predominantly), at the level of the signified.

A similar mechanism of resonance is used by the computer. Thus, the discovery of information kept in the computer's memory supposes: (a) its recognition by scanning / reconfiguration, after the introduction of the key-stimuli into the computer's "search-motor"; (b) the discovery of some similar configurations / forms / information stocked in the (magnetic) support of the computer's memory.

7. *The principle of binary processing data*: the "cerebral computer" operates with a system of information based on the presence or absence of the stimuli (light, sound, pressure, etc.), by an "On-Off", + /– , 1 / 0 mechanism. This mechanism permits the use of the formal logical laws, through some basic operators: conjunction, disjunction, negation, implication. With the help of these "technical" operators, the brain represents the necessary "hard" (signifier) for the "soft" of mind (signified), able to generate the signs used by human thinking (inferences) to construct all of its creative products.

The computer uses the "language of the electric signals", by the same "On-Off" mechanism. The presence-absence of electric stimuli – analogous to the presence or absence of the stimulating light, explained by BLT – is used to construct all the complex structures and information of the computational knowledge by a binary system.

2.2. On Language's Genesis: Contributions of Biophotonics

Using all these biophotonical principles, implicitly present in the computer's technology, we could:

a. justify and explain Pribram's essential hypothesis (1971, 35-52) that the brain's activity is of a holographic type; but, as it happens with technique, to generate and activate a biological hologram means to have a laser light and an instrument able to generate and use this light. It is here that our contributions could be very important, because BLT rationally describes this "technical" competence and performance of the biological structures;

b. offer an answer to the main problem formulated by Jackendoff: "How can we talk about what we see?" (Jackendoff, 1987, 90).

Taking into consideration the hypothetical principles and problems mentioned above, we could propose a biophotonic explanatory model for the word / language genesis. This model is based on the paradigm of iconicity, considered by Thomas A. Sebeok in his IH ("iconicity hypothesis") to be "the primary means by which bodily experience is transformed into a system of signs and meanings" (Danesi, 1998). In this spirit some other very recent contributions are formulated, such as: the "iconic theory of language genesis", the "motor theory of language" (Allott, 1988), etc. Although these theories are basically correct, they did not have the indispensable explanatory instruments of biophotonics. Using these instruments, our theoretical model is able to explain scientifically the stages of the brain activity, essentially involved in the genesis of the word, which we can only name in the present context:

- the genesis of the visual representations (mental holograms) as some stable "memory traces", drawn by some transparent "crystal liquid" structures; these "cerebral holograms" reproduce the properties of the referential object by a (bio)photonic / homeomorphic mechanism;
- the activation of the "mental luminous hologram" of an object, phenomenon, etc. in the absence of the referential (= process of representation), by controlling the volition and attention;
- the specific commands realized by the brain (transmitted from the central or peripherical brain sectors), in order to transfer the information comprised in the cerebral hologram (as a "photonic sign"), at the level of the verbal-kinaesthesic apparatus;
- each type of information reproduces a certain real property of the referential: colour, form, material, etc.;
- this information (hologram's vibrations) is transformed into mechanic (pressure) information at the level of the verbal- kinaesthetic apparatus, due to the same property of piezoelectricity of the liquid crystals;
- this type of information is simultaneously and synchronously (coherently) transferred at the level of the system of "linked resonators" (stomach and lung cavity, trachea, larynx, pharynx, mouth cavity), which defines the articulate-word as a complex type of SASER ("Sound Amplification by Stimulated Emission of Radiation").

*

The above proposed scenario is able to resolve the dispute between Cratylos and Hermogenes concerning the naturalist (iconic) and conventionalist (arbitrary) genesis of verbal language / words (Platon, 1978). Thus, we can say that:

- in the archaic phase of their use, words had a "natural" character, translating at the cerebral / vocal level some essential properties of the referential;
- in the modern phase of the language, words became predominantly arbitrary, being the result of some norms conventionally established in the process of social communication.

In this last category of language use, computational language is involved too, being characterized by some properties such as the following form Popa, (1972, 103-104): it is derived from natural language, it is highly formalized, using a limited number of symbols and rules able to describe a large informational content, it has a large area of utility, with a high degree of specialization and performance for accomplishing some cognitive functions, etc. Taking into consideration all these particularities, it is normal to formulate the following question: How can we measure the differentiation and correspondence established between artificial and natural language in the process of knowledge?

Trying to formulate an answer, we will observe that:

1. *Structurally* speaking, the machine language presupposes the great majority of the attributes present at the level of human language too, having as support the mechanism of resonance:

- the diffuse and specific sensibility (by dotting the computer with sensors sensible for any type of signals);
- the capacity to utilize some semantic codes, able to simulate the natural language of the human being;
- the possibility to stoke on a specific support and to actualize the main type of information / knowledge used by human language;
- the capacity to perform and to transmit information to man or to some other technological systems, etc.

2. *Functionally* speaking, the computational language is able to partially accomplish some of the functions of natural language, such as:

- to discover and interpret primary information concerning nature;
- to classify and name conventionally the discovered / interpreted realities;
- to generate an output able to overpass (qualitatively and quantitatively) the informational inputs;
- to enrich permanently the artificial intelligence system of signs.

3. *Structurally and functionally*, an integrative analogy between natural and artificial language is also possible. In this topic, we must take into consideration the dispute between the “inneist” and “constructivist” theory. From our point of view, it’s possible to consider that a mediator solution is appropriate (Stanciulescu, 1996): the genesis of the human language and knowledge must be both searched into the existence of:

- a “hard nucleus” (Chomsky), as a measure for the cognitive-linguistic competence (*a priori* / inborn knowledge), phylo-genetically gained by human species and hereditary transmitted to the individual;
- some psycho-logical capacities (Piaget), generated by the social-cultural medium *(a posteriori* / constructed knowledge), as a measure for the cognitive-linguistic performance, developed by human beings along their ontogenesis.

By referring only to the computer performances, as the instrument of the computational knowledge, we could say that it is analogous to the structure of any "macro-sign", namely:

- the hardware represents the physical support (signifier / technical "text") having in principle the same technical structure in the case of any computational system (being "innate", technologically implemented respectively);
- the software represents the coherent informational patterns (signified / artificial "discourse"), the "constructed" programs – functionally associated with the hardware – which is necessary for any operational purpose;
- the relationship between hard and soft is established by a "linguistic interface", represented by the computer's system of operation (DOS, Windows, etc.)., which generate the "computational signification / knowledge".

In conclusion, we could say that the possible analogies between the natural and artificial systems of knowledge impose the necessity of the "computational semiotics" proposed by H.N. Teodorescu (2000), able to formulate an answer for the obsessive question: "What are the future separating limits between computers and humans?"

3. The Computational Performance, a "Measure" for the Human Creative Competence!?

Formulated as an exclamation, the above assertion could suggest that between all the human technological performances, the realization of "artificial intelligence" / "computational knowledge" is certainly the most spectacular. As an interrogative the sentence may assert that the "artificial intelligence" is able to only partially simulate the creative mechanism of the human mind.

Each of the two alternatives supposes some specific considerations.

1. Synthetically, we could define the human natural knowledge as a bio-psycho-logical competence, historically constituted, which concerns man's capacity to operate with signs (with information, as signified, and energies-substances, as signifier), to reflect objective and subjective reality, to integrate the models resulting from reflection and signification into informational complexes constituting the universe of culture. Following the dimensions of this expansive definition, we could observe that all its dimensions are present in the case of "computational knowledge" too:

- the reaction to a certain category of stimuli (primary information, belonging to a "first system of signalizing", based on hard-soft relation), their intelligent and creative processing (by a "third system of signalizing");
- the ability to use a certain language (binary), in order to simulate many performances of natural language (as a "second system of signalizing), the computer being capable of "plurimediality" (the translation of the verbal information into a nonverbal one);
- the performance of auto-cognition represents a step made by last generation computers, able to evaluate their own functional states, their degree of solicitation, etc.;
- the performance to elaborate a "universe of cybernetic culture", integrating the experience of "learning automates", able to: auto-program, auto-regulate, auto-reproduce by "learning matrix", machines which "think" strategic and creative solutions for leading / decision / command, etc.

2. Based on this very last performance, creative "computational knowledge" is apparent. It is known that it's possible to learn and stimulate creativity, by activating intuitive or logical methods. The computer can use one of these in order to elaborate some heuristic strategies ("know-how", expert systems, etc.).

We could consider that all these strategies represent the interest nuclei of artificial intelligence, respectively: the problems of solving problems, by searching for solutions in the space of states (Barliba, 1990). In this respect (of heuristics, as art of invention, "*ars inveniendi*"), research of the computational epistemology (of "knowledge discovery" type) was developed having as a purpose the use of computers computers both for the discovery of new knowledge and for the development of creative thinking.

Even though a computer can accomplish all of these performances, it will ever remain only an instrument able of cognition projected from the interior of the human knowledge, even in the possible situation of its own reproduction. Of course, from a cybernetic point of view, the human being has a lot of "weaknesses" in comparison with the machine-computer: dysfunctional sensibility to some medium factors, partial fidelity of the memorized information, the reduced speed of the logical operation and the reduced number of simultaneous operations, the limited protection against the invasion of information, errors of evaluation in decision, action, etc.

But, despite such "weaknesses", the existence of the human being is characterized by some essential and strong BPLS (Bio-Psycho-Logical and

Social) features, already mentioned, able to make him / her a unique system of the world, synthetically taking into consideration that:

- biological life is the natural capacity of certain systems (vegetation animal, human) to receive and metamorphose the energy and information, passing by the cycle of birth, maturity, death; or, in biophotonic terms, life is the capacity of the above mentioned systems to permanently generate ultraweak (bio)luminescence, by metamorphosing the natural light from the physical state (non-coherent, non-polarized, etc.) to the state of "biolaser" light (with all the qualities of laser emission), and to entertain / metabolize it by permanent exchanges of energy and information with the external medium;
- *psychological attributes* concern the capacity to associate emotional-affective and volitional-optional states, and the possibility to auto-evaluate subjectively the own activity; or, biophotonically speaking, the psychic concerns the possibility to coherently use the "language of light" (bioluminescence), characterized by frequencies, intensities, modulation, etc., in order to sustain the bioelectromagnetic control of the emotional and volitional activity of the human being;
- *logical competence* represents the capacity to generate a mechanism of thinking and a history of knowledge, by creative reactions to the exogenous and/or endogenous stimuli, based on the activity of the cerebral support (hardware) and the denotative / connotative significations of human (non)verbal language (software); from a biophotonic perspective, it means to transform the "biological light" (bioelectromagnetic field) into an assembly of holograms representing ontological patterns (holograms) able to generate meanings / signified by their interconnections, and to express them by different creative signifiers;
- *social communication* defines the competence of the human being to be a *zoon politikon*, by interconnecting with his / her fellows for finding the best solutions for solving the problems of their common life; or, biophotonically, to correlate the emissions of human "living light" for determining a more "illuminated" / coherent frame of life, psycho-socio field, morphogenetic field, etc.

The synergy of all these resources is generating the human being's quality to be "face and alikeness" with his / her Creator, namely, to be a creative "being of light". But what about a computer? Could the human creator control and improve the attributes of the computer in order to be, as much as possible, his / her technological "face and alikeness"? We can answer this question creatively, appreciating that:

1. A computer cannot be a real living being, because life is still not generable by technological means. But, the machine's performance could

be improved by using some biological elements / components, such as a biological memory, by using the organic liquid crystal structures for greatly increasing the capacity of the computer's memory, or by realizing some biological circuits (optical bio-fibres) for transmitting the information in a (bio)photonic way, etc.

2. A possible computational application is to generate a bio-psychical interface between the human being and the computer, by using biosensors able to make the computer sensitive to human reactions and emotions, mental commands, etc. The computer's circuits – based on silicon crystals – are already sensitive enough to human bioelectromagnetic fields. But, using organic components, a computer could interact with its user much better.

3. The logic and creative competence of any computer could be increased in many ways, starting from the structural realization of some original "biological gates" until the use of methodological software – such as the "semiotic graph", able to describe the most complex semiosis and to determine the "critical road" / the best way for solving any type of problem (Stanciulescu, 2005, 160-167) and of combinatory programs able to generate original answers.

4. To improve the communication between human and computers, by using bioelectromagnetic technologies and channels, also means to realize a better "communication" among computers themselves, by realizing networks able to amplify the functional performances of the integrated machines, by interacting with their resources, memories, knowledge, etc. The internet is the explicit answer for such a performance, which could become a "living net", having the biophotonic holography, the mental commands and the biophotonic (laser type) interactions as the most performing results.

Permanently having in mind certain ethical limits, the time for connecting the bio-psycho-logical attributes of human beings with the art / logic of creation of computers is already coming. The rapprochement to such human attributes could become an aspiration for constructors of future computer generations. In fact, whether the machine will realize all these performances or not, it is already able to generate the necessary knowledge to essentially change man and the world.

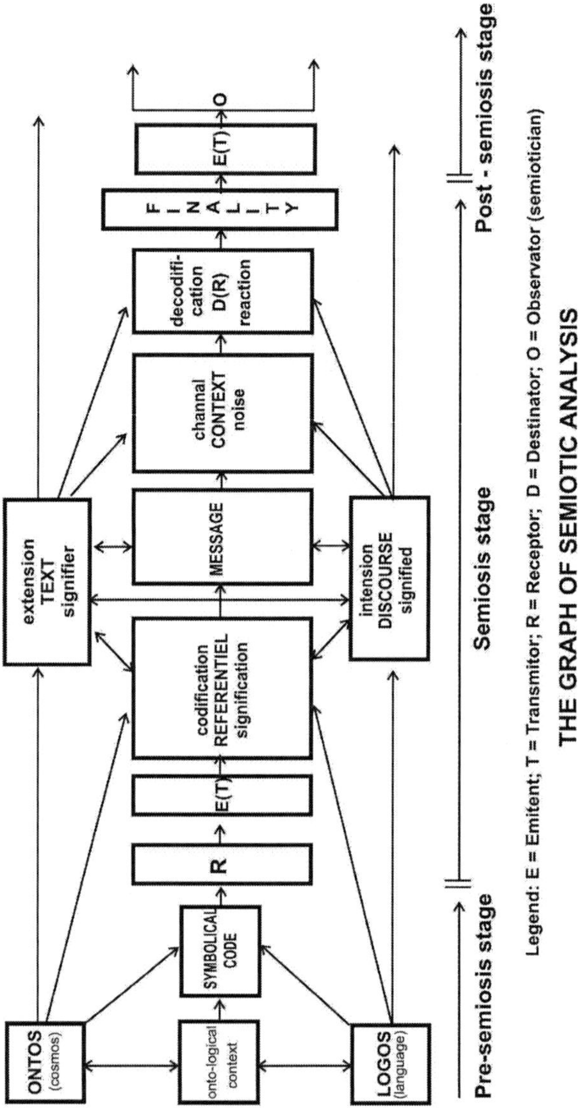

Figure 1. A schematic methodology of interaction between human beings and computers, by functionally connecting parameters such as: context of analysis (natural and cultural), symbolical code of the emitter and of the receptor, the content of the message and the transferring channel, the (de)codification process and the finality of the message, etc.

INSTEAD OF THE CONCLUDING THOUGHTS

VIOREL GULICIUC AND EMILIA GULICIUC

The philosophy of the beginning of this millennium seems to be torn between two extreme perspectives: the research from the cognitive sciences (especially in relation to artificial intelligence) and the post-modernism developed and recovered in trans-modernism.

The era prior to the Singularity, which Raymond Kurzweil defends (*The Singularity is Near*, 2005), is the era in which robotics, nanotechnology and genetic engineering arrived at the development stage; where problems with certain dimensions and philosophical allegiances are more and more frequent.

In this context, to take an example from the Artificial Intelligence domain, John McCarthy in *What has AI in Common with Philosophy?* (1995), goes further than Mario Bunge (see his interview with Jan-Kyrre Berg Olsen & Evan Selinger, in their book *Philosophy of Technology: 5 Questions,* Automatic Press / VIP, February 2007), saying the latest technologies and the intellectual ones rediscover an important philosophical problematic assigned traditionally to ontology, gnoseology or axiology. To the A.I. specialists a problem arises more often: the existence of a robot with intelligence comparable to man, with the capacity to learn from experience, for the creation of which it will be necessary to have a wide perspective over the world and human kind in order to serve its behaviour pattern.

Science and technology at the beginning of this millennium (including engineering), have reached a degree of mature understanding of the problems they confront themselves with, and that generates a considerable metaphysical pressure upon the classic topics.

Meanwhile, philosophy, in a world considerably more dominated by scientific referential, meets a symmetrical process: the increase of scientific rationality's pressure upon the issues that, traditionally, philosophy must answer to.

If we were to express mathematically the relationship between science and philosophy, then every increase in "philosophicality" is accompanied by a decrease in "scientificality", as every increase in scientific rationality accuracy is followed by a decrease in philosophical speculative rationality.

G. Deleuze and F. Guattari, in *What is philosophy?* (1996), believe that philosophy needs a non-philosophy to able to understand it, just as science or art need non-science and non-art.

Even if the theory of complexity is more and more frequent in science, philosophy, or art speeches, the idea that philosophy, art and science are descendents of chaos, remains under debate. Rationality, the common feature of these such heterogeneous domains, may be recovered, as it is the case for trans-modernism in a new formula: the "transversality" of its applicability, presented by Calvin Otto Schrag, in *The Resources of Rationality: A Response to the Postmodern Challenge (Studies in Continental Thought)*(1992).

In the societies of the former communist regime, more than in occidental communities, engineering seems to have gained in the past 50 years an almost religious dimension. That is why it is believable to consider that we will be able to use in our advantage, the distinction expressed by Mircea Eliade concerning the mythical and the historical time also, from the point of view of the philosophy of engineering; because every engineering work is essentially time consuming, and nowadays engineering has began to have a certain ritualised dimensions. This means that in everything that man does and has a meaning – ritual – (let's read: *from an engineering perspective*) as well as in everything that man thinks coherently – metaphysically – man aspires continuously towards unification, reintegration or completion.

As Paul Pierre Lévy (in *Collective Intelligence: Mankind's Emerging World in Cyberspace,* 1999) remarks, the opposition between techno-sciences and reflexive knowledge represents, nowadays, the main source of ignorance. The author considers philosophy, literature and culture, generally speaking, as carriers of a multi-millenary tradition of reflexion over knowledge, signs and communication. He observes, as well, that techniques participate in the reconstruction of the classical object of philosophers and humanists, because the creation and management of signs, the release and the transmission of knowledge, represents the newest and the most important concern of contemporary techniques and industry, of informatics and various domains.

Knowledge, as society's future, must be reconsidered. In this context, the problem of developing inter- and trans- disciplinary relationships becomes more valuable as well as the necessity to connect science and philosophy. The remarkable results obtained by the work of a team formed by humanists and scientists, in projects like *Arete* (University of Chicago) - especially while talking about the subproject *Redefining Wisdom*- demonstrates this.

By its origin, that already prefigures future (in the meaning confessed and undertaken by Martin Heidegger regarding his relationship to Edmund Husserl, as Jean Beaufret expresses it, in *Ein Vernunft aber bleibt stehts Zukunft*), philosophy remains a rational activity, permanently concerned, in one way or another, by favouring rationality. That is why it seems destined to a transversal rationality, as C. O. Schrag writes about.

As a continuous, endless interpretation, philosophy is also normative, since rationality expresses itself in principles, and principles allow us to interpret nature and mankind.

The same thing happens to engineering when it arrives at a degree of meta-theoretical maturity, thus generating the problems of the philosophy of technology, seen as specific features in the field of the philosophy of engineering.

If it is true that philosophy succeeds thanks to the betrayal of its servants (in the extent to which the interpreter, as a different translator, always betrays- situations expressing the result of the distance between the real and the unreal, the fact and the value; that is the risk in applying rules under rational signs), engineering, instead, succeeds somehow in a reverse way, thanks to the betrayals of its detractors.

In this context, engineering starts to put in discussion the very type of universality subsumed to human action, because *the engineer brakes the monolithically generic universality* (in which all x are reduced to y) of Plato-Aristotle type, in concrete universal non-generic accomplishments, in which universality is always incomplete, irreducible, but so alive and so present in front of us in the essence of the thing made "engineerically".

This is why we believe that the philosophy of engineering, as part of the larger field of technology philosophy, brings strong arguments in the favour of non-generic universality. The concerns of philosophy and technology have met a natural development in the domain of philosophy engineering.

With the entrance in the Knowledge Society, engineering sciences have gained not only an ethical dimension, that is becoming more and more substantial, but also a metaphysical one, since the philosophy itself has technical origins, if it is to believe Diogene Laertios, reminding us that both engineering as well as philosophy were created by Hephaistos.

The fact that in occidental societies the awareness of technology's and engineering's metaphysical dimensions, have been understood and accepted as research themes sometime before they were introduced in the societies from eastern Europe, is an expression of the different scale used to disseminate new information, rather than an expression of a theoretical immaturity[1] where creators of high intelligence, from both eastern and south Europe, philosophers, but also engineers, have understood to confront their perplexities – if we are to believe the definition given to philosophy by Jorge Luis Borges.

In Romanian philosophy a considerable preoccupation towards universality has existed in Mircea Eliade, Stephanne Lupasco, Lucian Blaga, Constantin Noica, Anton Dumitriu etc, maybe because technical creation has met in this country its best results though the work of János Bolyai, Traian Vuia, Henri Coandă, N. Paulescu, Gogu Constantinescu, Herman Oberth, George Emil Palade, etc.

That is why PHEADE 2009 came on a background of concerns for the reflexion upon the relationship between philosophy, science, technology and engineering, being ready to intervene into the contemporary debates regarding the mutations that science, technology and engineering will bring in the next years, changing not only reality, but also the perception of the frontiers between virtual and real[2].

Pierre Lévy, in *Les nouveaux outils du savoir* (Ed. Descartes), says that "in the cultural, semiotic and technical adventure of the humans that are building their own world and evolve with it, exists a certain type of dynamic curving of the transcendental: the present experience possibility condition is an anterior experience or one of a different rank".

Through the philosophy of engineering, both engineering as well as philosophy are profoundly involved in the probing of a new type of transcendental curving brought by the emergent technologies of the beginning of this new millennium.

That is why in this Era-Just-Before-the-Singularity the perplexities of our philosophies are not dissolved but required by the engineering.

---

[1] See PHEADE 2009 http://www.goldenideashome.com/pheade2009/

[2] Only around 75% of the papers presented at PHEADE 2009 have been accepted for this book. Among the papers non included, one will find: Interference between *Engineering and Humanities, The Senses and Sentiments of the Robots, Complexity thinking in Edgar Morin, Humanist implications of the Technology,theory of Information between Engineering and Philosophya.o.*This is clearly demonstrating the potential of the future editions of PHEADE conferences.

BIBLIOGRAPHY

Viorel Guliciuc

—. (2007) *Transcription of Discussions in Seminar*, 27 March 2007, on *Philosophy of Engineering*, at: http://www.raeng.org.uk/policy/philosophy/pdf/Transcript_of_Discussions_in_Seminar_on_27_March.pdf

—. (2007) "A Philosophy of Engineering Seminar: AI and IT: Where Engineering and Philosophy Meet", 11th July 2007, www.raeng.org.uk/events/pdf/Philosophy_flyer.pdf

—. (2008) Global *Catastrophic* Risk Conference (17-20 July 2008) website at: http://www.global-catastrophic-risks.com/index.html

Bao, O. (2007) "Engineering, Culture and Engineering Culture", in: Abstracts of the Workshop Philosophy & Engineering 2007, Delft University of Technology, October 29-31, 2007, http://www.illigal.uiuc.edu/web/wpe/files/2007/10/wpe2007abstracts.pdf , p. 76.

Chang, H. "The Myth of the Boiling Point", at: http://www.ucl.ac.uk/sts/chang/boiling/index.htm

Dear, P. (2001). "Science Studies as Epistemography", in: Labinger, Jay A. and Collins, H. M. (eds), *The One Culture? A Conversation About Science*, Chicago: University of Chicago Press, 2001

Dumitriu, A. (1977). History of Logic, Tunbridge Wells: Abacus Press, 1977

Goldberg, D. E. (2007). "Why Philosophy? Why Now? Engineering Responds to the Crisis of a Creative Era", in: *Abstracts of the Workshop Philosophy & Engineering 2007*, Delft University of Technology, Oct. 29-31, 2007, http://www.illigal.uiuc.edu/web/wpe/files/2007/10/wpe2007abstracts.pdf, pp. 35-36, the full paper in press.

Floridi, L. (2005) "Consciousness, Agents and the Knowledge Game", in: *Minds and Machines*, 2005, 15 (3-4) pp.415-444

Green, C. "What is Engineering Design?" at: http://www.ncedaust.org/index.php?select=63

Guizzardi, G. (2005). "Ontological Foundations for Structural Conceptual Models", CTIT PhD Thesis Series, No. 05-74, Telematica Instituut

Fundamental Research Series, No. 015 (TI/FRS/015), Enschede, 2005, at: http://www.telin.nl/index.cfm?type=doc&handle=55835&language=nl

Guliciuc, V. (2009a) "The Presupposition of Universality" in: Omelchenko, N. (ed.) (2009) The Human Being in Contemporary Philosophical Conceptions, Cambridge: Cambridge Scholar Publishing

Guliciuc, V. (2009b) "*How* do we need universities in a technological world?", in *Dialogue and Universalism*, 2009 …….

Laertios, D. The lives and Opinions of Eminent Philosophers, at: http://www.perseus.tufts.edu/hopper/text.jsp?doc=Perseus:text:1999.01.0259:page=3&highlight=philosophy

Joy, W. N. (2001). "Why the future doesn't need us" in: *Wired*, Issue 8.04, April, 2000, at: http://www.wired.com/wired/archive/8.04/joy_pr.html

Mitcham, C. (1998). 'The Importance of Philosophy to Engineering", in: *Teorema. Revista International de Filosofia*, vol. XVII/3 1998

Mitcham, C., and Mackay, R. (2007). "Comparing Approaches to the Philosophy of Engineering - Including the Linguistic Philosophical Approach", in: Abstracts of the Workshop Philosophy & Engineering 2007, Delft University of Technology, October 29-31, 2007, http://www.illigal.uiuc.edu/web/wpe/files/2007/10/wpe2007abstracts.pdf, pp.29-30, full paper *in press.*

McCarthy, N. (2008). *"Philosophy in the making",* in: *Ingenia*, 2006 March 26, pp. 47-51.

Monk, J. (2005) "The role of language in engineering competence", in: *International Conference on Engineering Education and Research*, 1-5 March 2005, Tainan, Taiwan, at: http://oro.open.ac.uk/5584/1/14-003.pdf

Parry, R. "Episteme and Techne", in: *Stanford Encycopedia of Philosophy*, at: http://plato.stanford.edu/entries/episteme-techne/

Prensky, M. (2009) "From Digital Immigrants and Digital Natives to Digital Wisdom", in *Innovate. Journal of online education*, Volume 5, Issue 3, February/March 2009, at: http://www.innovateonline.info/index.php?view=article&id=705&action=synopsis

Quinion, M. "Artefact versus Artifact", at: http://www.worldwidewords.org/qa/qa-art1.htm\

Vermaas, P. E.(2007). "What is the Use of Philosophy of Engineering? The case of the analysis of technical functions" in: *Abstracts of the Workshop* Philosophy & Engineering 2007, Delft University of Technology, Oct. 29-31, 2007, http://www.illigal.uiuc.edu/web/wpe/files/2007/10/wpe2007abstracts.pdf, pp. 25-26, the full paper in press.

Raymond Turner

Achinstein, P. (1983). *The Nature of Explanation.* New York: Oxford University Press.

Chalmers, A. (1999). *What is this thing called Science?* London: Open University Press.

Colburn. (2006). The Methodology of Computer Science. In Floridi. Oxford.

Colburn, T. (2000). *Philosophy and Computer Science.* New York: M.E. Sharpe. Inc.

Das, A. (1993). *The Special Theory of Relativity, A Mathematical Exposition.* New York: Springer.

Dym, C. L. (2009). *Engineering Design: A Project-Based Introduction.* Hoboken, NJ.: John Wiley & Sons.

Einstein, A. (1905). Zur Elektrodynamik bewegter Körper. *Annalen der Physik 17* , 17-89.

Ertas, A. J. (1998). *The Engineering Design Process, John Wiley and Sons, New.* New York: John Wiley and Sons.

Fetzer, J. (1988). Program Verification: the very Idea. *CACM* , 1048-1063.

Franssen. (2009). *The Philosophy of Technology.* Retrieved from Stanford Enclopedia of Philosophy: http://plato.stanford.edu/entries/technology/

Gould, S. (2002). *The Structure of Evolutionary Theory.* New York: Belknap Press.

Hyman, B. (1998.). *Fundamental of Engineering Design, Prentice Hall, New Jersey, 1998.* New Jersey: Prentice Hall.

Kohn. (2003). *Discussion of the Method.* Oxford: Oxford.

Kripke, S. (1982). *Wittgenstein on Rules and Private Language.* Harvard University Press.

Kuhn, T. (1970). *The structure of scientifc Revolutions.* Chicago: University of Chicago Press.

Lakatos, I. (1980). *The Methodology of Scientific Research programmes: Philosophical Papers Vol 1.* Cambridge.: Cambridge University Press.

McCarthy, N. (2006). Philosophy in the Making. *Ingenia, issue 26* .

Pfleeger, S. A. (2009). *Software Engineering.* Boston: Pearson.

Popper. (1972). *The Logic of Scientifc Discovery.* London: Hutchinson.

Rosenberg, A. (2000). *The Philosophy of Science: A Contemporary Introduction.* New York: Routledge.

Smith, M. (1975). *The Theory of Evolution.* Cambridge: Canto.

Wirth, N. (1974). On the Design of Programming Languages. *IEEE Trans. Software Eng.* , 386-393.

Lorenzo Magnani

Carse, Alisa L. (1999). "Facing up to moral perils: the virtue of care in bioethics" (1996) in *A Health Law Reader. An Interdisciplinary Approach*, ed. by Robinson, John H., Berry, Roberta M., and McDonnell, Kevin, (Durham, Carolina Academic Press, 1999), pp. 51-59.

Carstein Stahl, Bernd (2004). "Information, Ethics, and Computers. The problem of autonomous moral agent", *Minds and Machines* 14 (2004), 67-83.

Chopra, Samir and White, Laurence (2004). Artificial agents. Personhood in law and philosophy, in: *Proceedings of the 16th European Conference on Artificial Intelligence*, ed. by López de Mántaras, Ramon, and Saitta, Lorenza (Amsterdam: IOS Press, Amsterdam, 2004), pp. 635-639.

Clark, Andy. (2003). *Natural-Born Cyborgs. Minds, Technologies, and the Future of Human Intelligence*, Oxford and New York: Oxford University Press.

Dawkins, Richard (1989). *The Selfish Gene* (1976). Oxford: Oxford University Press, Oxford, second edition.

Floridi, Luciano, and Sanders, Jeff W. (2003). "The method of abstraction", in Negrotti, Massimo ed., *Yearbook of the Artificial. Nature, Culture, and Technology. Models in Contemporary Sciences.* Bern: Peter Lang

Floridi Luciano, and Sanders, Jeff W. (2004). "On the morality of artificial agents", *Minds and Machines* 14 (2004), 349-379.

Johnson, Deborah G. (2004). Integrating ethics and technology, European Conference Computing and Philosophy (E-CAP2004_ITALY), Abstract (Pavia, June 2-5, 2004, Italy).

Kant, Immanuel (1964). *Groundwork of the Metaphysics of Morals* [3d ed.] (New York: Harper & Row) reprint of the 1956 edited and translated by H.J. Paton (London, Hutchinson & Co., Ltd., third edition, 1964); originally published 1785.

Knorr Cetina, Karin (1999). *Epistemic Cultures. How the Sciences Make Knowledge.* Cambridge, MA: Harvard University Press.

Latour, Bruno (1999). *Pandora's Hope.* Cambridge, MA: Harvard University Press.

Magnani, Lorenzo (1997). *Ingegnerie della conoscenza. Introduzione alla filosofia computazionale.* Milan: Marcos y Marcos.

—. (2001). *Abduction, Reason, and Science. Processes of Discovery and Explanation.* New York: Kluwer Academic/Plenum Publishers.

Magnani, Lorenzo (2007). *Morality in a Technological World. Knowledge as Duty.* Cambridge: Cambridge University Press.
Mill, John Stuart (1966). *On Liberty* (1859), in: J.S. Mill, *On Liberty, Representative Government, The Subjection of Women*, Oxford: Oxford University Press, twelfth edition, pp. 5-141.
Powers, Thomas M. (2004). Intentionality and moral agency in computers, European Conference Computing and Philosophy (E-CAP2004_ITALY), Abstract. Pavia, June 2-5, 2004, Italy.
Rachels, James (1999). *The Elements of Moral Philosophy*, McGraw Hill College, Boston Burr Ridge, IL
Shelley, Cameron (2004). Disanalogies and the validity of animal models, International Conference Model Based Reasoning in Science and Engineering (MBR04), Abstract, Pavia, December, 16-18, 2004, Italy.
Singer, Peter (1974). "All animals are equal", in *Philosophic Exchange* 1(5) (1974), 243-257.
Urban Walker, Margaret (1996). Feminist scepticism, authority and transparency. In *Moral Knowledge? New Readings in Moral Epistemology*, ed. by Sinnott-Armstrong, Walter and Timmons, Mark, eds. (Oxford: Oxford University Press), pp. 267-292.

Peter Boltuc

Baars, B. J. *A Cognitive Theory of Consciousness.* Cambridge University Press, 1988
Boltuc, N.; Boltuc, P. "Replication of the Hard Problem of Consciousness in AI and Bio-AI: An Early Conceptual Framework" in: Chella 2007
Boltuc, P. "Introduction to Complementary Philosophy", *Colloquia Communia* (Pol.) 3-4 (1987) 259-64
Boltuc, P. 1990 "Person as a Locus of Permanence: Towards Albert Shalom's Metaphysics." *Dialectics and Humanism* 2: 213-34
Boltuc, P. *The Problem of Subject and Object in English-Language Philosophy* (Pol.) – unpublished dissertation
Boltuc, P. 1998a. "Reductionism and Qualia", *Epistemologia* 4 111_130.
—. 1998b. "Qualia, Robots and Complementarity of Subject and Object, World Congress of Philosophy", Boston http://www.bu.edu/wcp/Papers/Mind/MindBolt.htm
Boltuc, P. 2007. "Does epistemic subjectivity have moral import?" *Analiza i Egzystencja* 6 5-24.
Boltuc, P. 2008. "Does Epistemic Subjectivity Have Moral Impact?" (Pol.) *Analiza i Egzystencja* 6: 5-23

Boltuc, P. 2009. "The Philosophical Issue in Machine Consciousness", *International Journal of Machine Consciousness* Vol. 1, No. 1

Boltuc, P. 2010a . "The Engineering Thesis in Machine Consciousness"

Boltuc, P..2010. "Sloman and H-Consciousness" *International Journal of Machine Consciosuness* Vol. 2, No. 3

Chalmers, D. 2003. "Consciousness and Its Place in Nature," in: Stich, S. Warfield, F. eds. Blackwell Guide to Philosophy of Mind. Blackwell

Chella, A.; Manzotti, R/ *AI and Consciousness: Theoretical Foundations and Current Approaches* AAAI Press. Merlo Park, CA, 2007

Dennett, D. 2005 *Sweet Dreams. Philosophical Obstacles to Science of Consciousness*, MIT Press

Floridi, L.; Artificial Companions and their Philosophical Challenges,*E-Mentor* 5/2007 http://www.e-mentor.edu.pl/artykul_v2.php?numer=22&id=498

Franklin, S. Baars, B. J. Ramamurthy, U. 2008 "A phenomenally conscious robot?", APA Newsletter on Philosophy and Computers, Vol. 1.

Haikonen, P. O. A. 2009 "Qualia and Conscious Machines", *International Journal of Machine Consciousness* 225-234

Johnson, D. Miller, K. 2009 *Computer Ethics: Analyzing Information Technology*, 4th Ed. Prentice-Hall

Moor, J. 2007 "Taking the Intentional Stance Towards Robot Ethics", *APA Newsletter on Philosophy and Computers* 06/02, 14-17

Nagel, T. 1986 *The View from Nowhere*, Oxford University Press

Siemek, M. 1984 *Die Idee des Transcendentalismusbei Fichte und Kant* F. Meiner, Hamburg

Sloman, A. "Why some machines may need qualia and how they can have them", in: Chella ed.

Turing, A. 1950*"Computing Machinery and Intelligence", Mind LIX (236): 433–460.*

Colin Schmidt

Archer, N., and Head, M., and Wollersheim, J. and Yuan, Y. (1996). "Investigation of Voice and Text Output Modes with Abstraction in a Computer Interface", in *Interacting with Computers,* vol. 8 n° 4, Amsterdam: Elsevier Science, p. 323-345.

Brooks, R. (2002). *Robot: The Future of Flesh and Machines*, London: The Penguin Press.

—. (1991). "Intelligent Room without Representation", in *Artificial Intelligence*, n° 47, 139-159.

Brooks R. & Stein L. (1994), "Building Brains for Bodies", in *Autonomous Robots*, n° 1, p. 7-25.

Button, G. and Coulter, J. and Lee, J. and Sharrock, W. (1995). *Computers, Minds and Conduct.* Oxford: Polity-Blackwell.

Card, S., and Moran, T. and Newell, A. (1983). *The Psychology of Human-Computer Interaction*, Hillsdale: Lawrence Erlbaum Associates.

Carroll, J., and Moran, T. (1991), Special Issue of *Human-Computer Interaction* on 'Design Rationale', vol. 6 n° 3 & 4, Hillsdale: Lawrence Erlbaum Associates.

Dahlbom B. (1993). *Dennett and his Critics. Demystifying Mind.* Oxford: Basil Blackwell.

Dascal, M. (1992). "Why does Language Matter to Artificial Intelligence? ", in *Mind and Machines*, 2, Dordrecht: Kluwer Academic Publishers, p. 145-174.

Dautenhahn, K. (1997). "The Role of Interactive Conceptions of Intelligence and Life in Cognitive Technology", in *Proceedings of the Second International Conference on Cognitive Technology*, Aug. 25-28, Aizu Japan, IEEE p. 33-43.

Dautenhahn, K. and Nehaniv, C. L. (2002). *The Agent-Based Perspective on Imitation, Imitation in Animals and Artifacts*, MIT Press, 2002.

Dennett, D. (1996), [*Kinds of Minds*] tr. in French; *La diversité des esprits : Une approche de la conscience*, Paris : Hachette.

—. (1987). *The Intentional Stance,* Cambridge: Bradford/The MIT Press.

—. (1971) "Intentional Systems", in *Journal of Philosophy*, n° 8, p. 87-105

Dreyfus, H.L. (1972). *What Computers* Still *Can't Do. A Critique of Artificial Reason,* Cambridge: The MIT Press.

Fogg, J. and Nass, C. (1997). "Silicon Sycophants: The Effects of Computers that Flatter", in *International Journal of Human-Computer Studies*, May issue.

Gazdar, J. (1979*). Pragmatics, Implicature, Presupposition and Logical Form*, New York: Academic Press.

Grice, P. (1975). "Logic and Conversation", COLE P. & MORGAN J. *Syntax and Semantics: Speech Acts*, vol n° 3, New York: Academic Press.

Hofstadter, D. (1979). *Godel, Escher, Bach: an Eternal Golden Braid,* New York: Basic Books.

Jacques, F. (1990). «De "On Denoting" de B. Russell à "On Referring" de P.F. Strawson, l'avenir d'un paradigme», in *Hermès VII: Bertrand Russell, de la logique à la politique*. Paris: Editions du CNRS, p. 91-100.

—. (1979). *Dialogiques, Recherches logiques sur le dialogue*, Paris: PUF.
—. (1976). "A Terminological Note about Reference", in Ryle, G., *Contemporary Aspects of Philosophy*, London: Oriel Press.
Mey, J. and Tamura, H. (1994). "Barriers to Communication in a Computer Age", in *AI & Society*, n° 6 Springer, p. 62-77.
Milner, P.M. (1999). *The Autonomous Brain: A Neural Theory of Attention and Learning,* Mahwah: Lawrence Earlbaum Associates.
Norman, D. (2001). "How Might Humans Interact with Robots? Human-Robot Interaction and the Laws of Robotology", keynote address, in *The DARPA/NSF Conference on Human-Robot Interaction*, San Luis Obispo CA, September. www.jnd.org/dn.mss/Humans_and_Robots.html
Peiris, D., and Gregor, P and Alm, N. (in print). "The Effects of Simulating Human Conversational Style in a Computer-based Interview", in *Interacting with Computers*, Amsterdam: Elsevier Science.
Putnam, H. (1981). "Brains in a Vat" and "A Problem with Reference" in *Reason, Truth and History,* Cambridge: Cambridge University Press.
—. (1975). *Mind, Language and Reality*. vol. 2, Cambridge: Cambridge University Press.
—. (1960). "Minds and Machines", Hook S., *Dimensions of Mind*, London: Collier-Macmillan.
Quine, W.V.O. (1960). *Word & Object.* Cambridge: The MIT Press.
Scassellati, B. (2001). *Foundations for a Theory of Mind for a Humanoid Robot*, Ph.D. Dissertation, May 6: Massachussetts Institute of Technology.
Schmidt, C.T.A. (2008). "Redesigning Man?", in Vermaas, Pieter E. and Kroes, Peter and Light, Andrew and Moore, Steven A. *(eds.) Philosophy and Design: From Engineering to Architecture,* Dordrecht: Springer Science, pp. 199-206.
—. (2006). "A Relational Stance in the Philosophy of Artificial Intelligence", *in* Magnani L. (ed.), *Computing and Philosophy*, Pavia: Associated International Academic Publishers
Schmidt, C.T.A. and Kraemer, F. (2006). "Robots, Dennett and the Autonomous. A Terminological Investigation", in *Minds and Machines. Journal for Artificial Intelligence, Philosophy and Cognitive Science*, Vol. 20 N° 1, Springer.
Schmidt, C.T. 2004. "Let Me Introduce You to My Non-Agent", *in* Schaub, H., and Detje, F. and Brüggemann, U. (eds.) *The Logic of Artificial Life*, Berlin: Aka-Verlag/IOS Press, p. 122-127.

Schmidt C. (2001). *L'esprit et la machine, une communauté conceptrice à l'œuvre. Distinctions pour un discours supérieur* (in French) [*Mind and Machine, Concepts for a Design Community at Work*], Lille: Presses universitaires de Septentrion.

—. (1997). "Pragmatically Pristine, the Dialogical Cause", 'Open Peer Community Invited Commentary', on Mele, A., "Real Self-deception", *Behavioral and Brain Sciences*, 20:1, Cambridge: Cambridge University Press

—. (1997). "The Systemics of Dialogism: On the Prevalence of the Self in HCI Design", in *Journal of the American Society for Information Science,* vol. 48 n° 11 November, New York: John Wiley & Sons.

Shannon, C. and Weaver, W. (1949). *The Mathematical Theory of Communication*, Urbana: University of Illinois Press.

Searle, J. (1975). *Speech Acts*, Cambridge: Cambridge University Press.

—. (1969). "Indirect Speech Acts", Cole, P. and Morgan, J. *Syntax and Semantics: Speech Acts*, vol n° 3, New York: Academic Press.

Shotter, J., (1997). "AI and the Dialogical", in *American Behavioral Scientist,* vol 40, n° 6, Special Issue edited by Harré. R. on "Computation and the Mind", p. 813-828.

Sperber, D. and Wilson, D J. (1986). *Relevance: Communication and Communication*, Oxford: Blackwell.

Strawson, P. (1950). "On Referring", *Mind*, vol. LIX July n°235.

Suchman, L. (1987). *Plans and Situated Actions: The Problem of Human-Machine Communication*, Cambridge: Cambridge University Press.

Turing, A. (1950). "Computing Machinery and Intelligence", in *Mind*, vol. LIX n° 236.

Turkle, S. (1984). *The Second Self, Computers and the Human Spirit.* New York: Simon & Schuster.

Vernant D. (1996). «Intelligence de la machine et sa capacité dialogique» [in French] [Machine Intelligence and the Dialogical Capacity of the Machine]; Rialle, V. and Fisette, D., *Penser l'ésprit: des sciences de la cognition à une philosophie cognitive*, Grenoble: Presses universitaires de Grenoble.

Watanabe, T. (2002). "InterActor: Speech-Driven Embodied Interactive Actor", in *Proceedings of the IEEE International Workshop on Robot and Human Interactive Communication* RO-MAN 2002, Sept. 25-27 2002, Berlin Germany: IEEE, p. 430-435.

Weng, J., and McClelland, J., and Pentland, A., and Sporns, O., and Stockman, I., and Sur, M. and Thelen, E. (2001). "Automous Mental Development by Robots and Animals", in *Science Magazine,* Vol. 291

N° 5504, The American Association for the Advancement of Science (AAAS), pp. 599-600.

Wittgenstein, L. (1953). *Philosophical Investigations*, Oxford: Blackwell.

Zlatev, J. (2001). "The Epigenesis of Meaning in Human Beings, and Possibly in Robots", in *Minds and Machines* n° 11, Kluwer, p. 155-195

Kuruvilla Pandikattu

Fukuyama, F. (1999). "Second Thoughts: The Last Man in a Bottle" in *The National Interest*, Summer

Hayes, R. (2009). "The Politics of Genetically Engineered Humans" on the website of the Organic Consumers' Association. Accessed at: http://www.organicconsumers.org/Patent/gehumans.cfm, February 2009.

Kant, Imm. (2004) *Critique of Practical Reason,* Thomas Kingsmill Abbott (tr), Barnes & Noble.

Linkinghub (2008) linkinghub.elsevier.com/retrieve/pii/S1369848602000262

Obama, B. (2009). (Washington: PTI news March 10, 2009) See http://www.ptinews.com/pti%5Cptisite.nsf/0/89289CC8522872D3652 5757500390EFB?OpenDocument

Pandikattu, K. (2004) *Tamas: There Are Many Alternative Stories.* Pune: Jnanam

Pence, G. (1998). *Who's Afraid of Human Cloning?* New York: Roman & Littlefield

Ricoeur, P. (2008) quoted from kurusj.googlepages.com/spiritualinformation/

Silver, L. L. (1977) *Remaking Eden: How Cloning and Beyond Will Change the Human Family,* New York: Avon Books

TOI (2008) Accessed from http://timesofindia.indiatimes.com/Health--Science/Science/Vellore-research-centre-makes-stem-cell-breakthrough/articleshow/4242720.cms. October 2008.

—. (2009) "Diagnosis Made Easy, with DNA Mapping," Times of India, Oct 16, 2009, Patna, p. 1.

Transhumanism (2009), at http://www.transhumanism.org/resources/faq.html.

Watson, G. S, and John Campbell, eds. (2000) *Engineering the Human Germline.* New York: Oxford University Press, 2000

Ionut Isac

Fellows, R. (1995), *Philosophy and Technology*, Cambridge: C, U, Press

Lawson, C. (2008), "An Ontology of Technology: Artefacts, Relations and Functions" in *Techné: Research in Philosophy and Technology*, 12 (1). http://scholar.lib.vt.edu/ejournals/SPT/v12n1/lawson.html, accessed01.05.2008.

Vries, M.J. de (2008) "Gilbert Simondom and the Dual Nature of Technical Artifacts." *Techné: Research in Philosophy and Technology*, 12 (1). http://scholar.lib.vt.edu/ejournals/SPT/v12n1/devries.html

Kaplan, F. (2009), "Intégration, incorporation, interface – l'évolution des systèmes techniques" in *Association F. Gonseth, CAHIERS de l'Institut de la Méthode*, 31 (Janvier 2009), St.-Imier, Switzerland

Cerqui, D. (2009), "Derrière la technique, l'humain" in *Association F. Gonseth, CAHIERS de l'Institut de la Méthode*, 31 (Janvier 2009), St.-Imier, Switzerland

Nicolescu, B. (2003), "Le tiers et le sacré" in *Le Sacré aujourd'hui, precedé de «Hommage à Michel Camus»,* Monaco: Éditions du Rocher (directrion de Basarab Nicolescu)

Sorin-Tudor Maxim

Callicott J. B. (2000) "Animal Liberation: A Triangular Affair" in *Environmental Ethics*, 2, 1980

Drucker, P. F., (1983) Managementul viitorului, [in Romanian] {The Management of the Future] Bucharest: ASAB Publishing House

Habermas, J. (1983) *Moralbewußtsein und Kommunikatives Handeln*, 1983, Suhrkamp Verlag Frankfurt am Main

Lipovetsky, G.(1993) *Le Crépuscule du devoir: L'ethique indolore des nouveax temps démocratiques*, Paris: Editions Gallimard

Morar. V. (2006). Moralităţi elementare. [In Romanian] [Elementary moralities] 2nd edition, Bucureşti: Paideia

Toffler, A., and Toffler, H. (2006) *Avuţia în mişcare* [in Romanian] [transl of. Revolutionary Wealth: How It Will Be Created and How It Will Change Our Lives] Bucharest: Antet

Bogdan Popoveniuc

Blackmore, S. (1999). The Meme Machine. Oxford: Oxford Univ. Press

Blaga, L. (1947). Despre conştiinţa filosofică [in Romanian] [Concerning Philosophical Consciousness] Cluj-Napoca: Lito-Schildkraut

Brooks, H. (1980). "Technology, Evolution, and Purpose", Daedalus 109, Winter

Dawkins, R. (1989). The Selfish Gene. Oxford: Oxford University Press

Ellul, J. (1983). "The Technological Order" in Mitcham, Carl and Mackey, Robert, *Philosophy and Technology: Readings in the Philosophical problems of Technology*, New York: Free Press, pp. 86-107

—. (1964). The Technological Society. New York: Vintage Books

Feenberg, A. (1999). *Questioning Technology*. NY & London: Routledge

Habermas, J. (1972). *Knowledge and Human Interests*. London: Heinemann

Heidegger, M. (1954/2003). „The Question Concerning Technology" in Scharff, Robert C. and Dusek, Val (*eds.*) *Philosophy of technology: the technological condition: an anthology*, Wiley-Blackwell

Jin, Z. (2005). *Global technological change: from hard technology to soft technology*. Intellect Books

Maturana, H. R. (1970). "Biology of Cognition", *in* Maturana, H. R. and Varela, F. J. (1980), *Autopoiesis and Cognition: The Realization of the Living*, Dordecht: D. Reidel Publishing Co., pp. 5-58

Maxwell, N. (2007). *From Knowledge to Wisdom: A Revolution for Science and the Humanities*. London: Pentire Press

Merrill, R. S. (1968). "Technology" in *International Encyclopedia of the Social Sciences*, vol. 15, ed. David L. Sills New York: Macmillan

Mitcham, C. and Mackey, R. (1983). *Philosophy and Technology: Readings in the Philosophical problems of Technology*. New York: Free Press

Popoveniuc, B. (2008). *Curs de Antropologie filosofică* [Ro] (Philosophical Anthropology Handnotes), Suceava: "Stefan cel Mare" University Press

Ricoeur, P. (1984). *Time and Narrative*. Vol. I Chicago: Chicago Univ. Press

Skolimowski, H. (1983). „The Structure of Thinking in Technology" *in* Mitcham, C. and Mackey, R, *Philosophy and Technology: Readings in the Philosophical problems of Technology*, New York: Free Press, 1983, pp. 42-49

Stiegler, B. (1998). *Technics and Time: The fault of Epimetheus*. Stanford: Stanford University Press

Emilia Guliciuc

Codoban, A. *Filosofie [Philosophy]*, http://www.scribd.com/doc/9197769/Aurel-Codoban-Filosofie/

(DEX) *Dicționarul Explicativ al Limbii Române [Romanian Language Explicative Dictionary]*, http://dexonline.ro/

Mirea, G. A., *Charmides – despre înțelepciune*, http://www.bookblog.ro/x-gabriel-mirea/charmides-despre-intelepciune/

(NODEX) *Noul Dicţionar Explicativ al Limbii Române [New Explicative Dictionary of the Romanian Language]*, http://www.webdex.ro/online/noul_dictionar_explicativ_al_limbii_romane/

Pană, L. (2000). *Filosofia culturii tehnice [in Romanian][Philosophy of the Technical Culture]*, Bucureşti: Editura Tehnică

Catalina Raducu & Mihai Floroiu

Berne Convention for the Protection of Literary and Artistic Works, http://www.wipo.int/treaties/en/ip/berne/trtdocs_wo001.html

Convention Establishing the World Intellectual Property Organization, http://www.wipo.int/treaties/en/convention/trtdocs_wo029.html

Paris Convention for the Protection of Industrial Property, http://www.wipo.int/treaties/en/ip/paris/trtdocs_wo020.html

Understanding Industrial Property, WIPO Publication No. 895(E), http://www.wipo.int

The Council Directive 91/250/EEC of 14 May 1991 on the legal protection of computer programs (*Official Journal L 122 , 17/05/1991 P. 0042 – 0046)*, http://eurlex.europa.eu/LexUriServ/LexUriServ.do?uri=CELEX:31991L0250:EN:HTML

Digital Millennium Copyright Act, Section 1201, f (1)-(4), http://www.eff.org/IP/DMCA/hr2281_dmca_law_19981020_pl105-304.html

Nicolescu, B. (2002) *Noi, particula şi lumea*, Iasi: Polirom

Kaner, C. (1998). "The problem of reverse engineering." *Software QA*, Vol. 5, #5, 1998

Samuelson, P. and Suzanne, Scotchmer, *The Law and Economics of Reverse Engineering*, The Yale Law Journal (Vol. 111: 1582-1663), available at http://www.yalelawjournal.org/pdf/111-7/SamuelsonFINAL.pdf

Florin Munteanu

Chalmers, D. (1996). *The Conscious Mind*. Oxford: Oxford University Press

Drăgănescu, M. (1985). *Ortofizica* [in Romanian] [Ortophysics]. Bucuresti: Editura Stiintifica si Enciclopedica,

—. —. (1979). *Inelul lumii materiale* [in Romanian] [The Ring of the Material World]. Bucuresti: Editura Politica,

—. (2003). *Societatea conştiinţei* [in Romanian] [The Society of Consciousness], Bucuresti: Editura Academiei Române

Erdi, P. (2008). *Complexity Explained.* Berlin: Springer-Verlag

Hoffmeyer, J. (2008). *Biosemiotics*. Place: Scraton University Press

Larche, J.-C. (2001). *Terapeutica bolilor spirituale* [in Romanian] [Therapeutics of the Spiritual Ilness]. *Bucuresti:* Editura Sophia

Maiorescu, T. G. (2001). *Introducere* în *ecosofie* [in Romanian] [Introduction in ecosophy]. Bucuresti: Fundaţia Europeană E.C.E.

Nicolescu, B. (2008). *Transdisciplinarity-Theory and Practice.* Cresskill: Hampton Press

Săhleanu, V. (1973). *Eseu de biologie informaţională* .[in Romanian] [An Essay in the Informational Biology]. Bucuresti: Editura Stiintifica si Enciclopedica

Boris Plahteanu

Belous, V. and Plahteanu, B. (2005). *Fundamentele creaţiei tehnice* [in Romanian] [Foundations of the Technical Creativity]. Iaşi: Performantica.

Plahteanu, B. (1999). *Ingineria valorii şi performanţa în creaţia tehnică* [in Romanian][The Engineering of Values and the Performance in the Technical Creativity]. Iaşi: Performantica.

Plahteanu, B. and Pleşu, Gh. and Călătoru, V. and Hanganu, Şt. (2001). *Concepţia şi proiectarea creativă a maşinilor unelte* [in Romanian][The Understanding and the Creative Design of the Industrial Machines]. Iaşi: Performantica.

Plahteanu, B. (2004), *Fondul de efecte fizice pentru inventatori* [in Romanian][The Physical Effects Database for Inventors], Iaşi: Performantica.

—. (2001)., "About Certain Making – Decisions Methods And means in the design of the integrated manufacturing systems", in *Buletinul Institutului Politehnic din Iaşi*, Universitatea Tehnică "Gheorghe Asachi", Iaşi, Tomul XLVII (LI), Supliment, 2001, Secţia construcţii de maşini.

Chenevier, J. (1980). "Innover pour survivre", in : *Science et vie. L'invention et l'innovation*, Nomero hors serials trimestrial .,Dec. 1980, p.42-45.

Maliţa, M., (1999) *Mileniul III - Atelier de idei şi proiecte ale civilizaţiei globale*, [in Romanian][Millennium III - Workshop of Ideas and Projects for European Integration and Global Civilisation], Academia

Română Fundaţia Universitară a Mării Negre, nr 2. p.13-19, 1999, Ed. Curentul , Bucureşti.
Janet, P. (1909). *Cours au Collège de France, 1909-1910: Analyse des tendances et lois de leur réalisation.* L'Harmattan, Paris, ré-édition, 2004.

Virgil Moldovan

Bunge M. (1984). *Ştiinţă şi filosofie.* [in Romanian][Science and Phylosophy] Bucuresti:Editura politică,pp 370 –441.
Drăgănescu, M.(1996).*L'Universalite Ontologique de l'Information.* [in French]. Bucuresti: Editura Academiei, http://www. racai.ro books/ drăgănescu/, chap.
10 html.Press, pp.19-38.
Fellows, R.(1995).*Philosophy and Technology*, Cambridge: Cambridge University Press.
Heidegger, M. (1982). "Intrebare privitoare la tehnică" [in Romanian][Question on Technique], in *Originea operei de artă* [in Romanian][Origin of Artwork].Bucureşti:Editura Univers.
Mitcham, C. (1994).Thinking trough Technology. Chicago: University of Chicago,
Puech, M.(2008). *Homo Sapiens Technologicus. Contemporary Philosophy of Technology. Contemporary Philosophy of Wisdom,* (in French*),* Paris:Edition Le Pommier, pp.7, 8, 314-324
Moldovan,V.(2008)."The Paradigms of Technology" in *Acta Tehnica Napocensis* no. 51, volIII, Universitatea Tehnica din Cluj-Napoca, Cluj-Napoca

Cristian Suteanu

Bateson, Gregory. 1972. *Steps to an ecology of mind.* New York: Ballantine.
Bohm, David. 1980. *Wholeness and the implicate order*. London: Routledge & Kegan Paul.
Boisot, Max H. 1995. *Information space: a framework for learning in organizations, institutions and culture*. London: Routledge.
Bugeja, Michael. 2005. *Interpersonal divide: the search for community in a technological age.* Oxford: Oxford University Press.
Connor, S. 2001. The decomposing voice of postmodern music. *New Literary History* 32:467-483.

Cook, T. 2001. Archival science and postmodernism: new formulations for old concepts. *Archival Science* 1:3-24.

Corbett, Julia B. 2006. *Communicating nature: how we create and understand environmental messages*. Washington: Island Press.

Depoe, Stephen P., ed. 2006. *The environmental communication yearbook.* London: Lawrence Erlbaum Associates Publishers.

Draganescu, Mihai. 1979. *The Depths of Existence.* http://www.racai.ro/books/doe/ (accessed July 5, 2009).

Duncan, T.L. and J.S. Semura. 2004. The deep physics behind the second law: information and energy as independent forms of bookkeeping. *Entropy* 6:21–29.

Fearn-Banks, Kathleen. 2007. *Crisis communications: a casebook approach.* London: Lawrence Erlbaum Associates Publishers.

Floridi, L. 2007. A look into the future impact of ICT on our lives. *The Information Society*, 23, 1, 59-64.

Foltz, Bruce V., and Robert Frodeman, eds. 2004. *Rethinking nature: essays in environmental philosophy*. Bloomington: Indiana University Press.

Georgescu-Roegen, Nicholas. 1971. *The entropy law and the economic process*. Cambridge, Massachusetts: Harvard University Press.

Kofler, J. and A. Zeilinger. 2006. Le paradoxe du chat de Schrödinger. *Sciences et Avenir Hors-Série* 148(54):11-12.

Landauer, R. 1996. The physical nature of information. *Physics Letters A* 217:188-193.

Levine, Robert. 1997. *A geography of time*. New York: Basic Books.

Mandelbrot, Benoit B. 1975. *Les objets fractals: forme, hasard et dimension*. Paris: Masson.

Mandelbrot, B.B. 1985. Self-affinity and fractal dimension. *Physica Scripta*, 32, 257-260.

McLuhan, Marshall. 1964. *Understanding media*. New York: McGraw-Hill.

Mitchell, Bruce. 2002. *Resource and environmental management*. Upper Saddle River: Prentice Hall.

Mokros, H.B. and B.D. Ruben. 1991. Understanding the communication-information relationship: levels of information and contexts of availabilities. *Knowledge: Creation, Diffusion, Utilization* 12(4):373-388.

Roff S.S. 2005. The return of the armchair scholar. *Journal of Scholarly Publishing* 1:49-57

Shannon, C. 1953. Abstract to The lattice theory of information. *IEEE Transactions on Information Theory* 1(1):105.

Suteanu, C. 2000. On the processes of meaning in geodynamics. *Revue Roumaine de Géophysique* 44:3-35.
—. 2005. Complexity, science and the public - the geography of a new interpretation. *Theory, Culture and Society* 22(5):113-140.
—. 2007. You cannot swim in foam: information flooding and information dilution in processes of change. *International Journal of the Humanities* 5(5):69-76.
Urry, John. 2003. *Global complexity*. Cambridge: Polity.
von Bayer, Hans Christian. 2004. *Information: the new language of science*. Cambridge, Mass.: Harvard University Press.

Dan Milici

Casti, J.L., and Karlqvist, A. (1991). *Beyond Belief. Randomness, Prediction and Explanation in Science.* Boston: CRC Press, Boston
Davies, P. (1989). *The New Physics*. Cambridge: Cambridge University Press
Haken, H. (1988). *Information and Self-Organization. A Macroscopic approach to Complex Systems*. Berlin: Springer-Verlag
Milici, D., and Crețu, M., and Milici, M., and Bobric, C. (2001). "Measurement as an informational instrument", in *Proceedings* of the International Metrology Conference celebrating the 50 anniversary of the Romanian National Institute of Metrology
Milici, D., and Crețu, M., and Milici, M., and Bobric, C. (2001). "Problems of the classical physics concepts reflected in metrology", in *Proceedings* of the International Metrology Conference celebrating the 50 anniversary of the Romanian National Institute of Metrology
Penrose, R. (1989). *The Emperor's New Mind – Concerning Computers, Minds, and the Laws of Physics*. Oxford: Oxford University Press
Zurek, W.H. (1990). *Compexity, Entropy and Physics of Information,* Santa Fe Institute Studies in the Science and Complexity, vol VIII
Astrov I., and Tatarly S. (2006) "Simulation of Two-Rate Neural Network Control for Stochastic Model", in *Advances in Electrical and Computer Engineering*, No 1/2006.
Rîşneac B., and Milici L.D., and Rață E. (2004). *Utilizarea tehnicii de calcul în evaluarea performanțelor.* Braşov: Editura Universității Transilvania
Yaneer, Bar-Yam, Applications of Complex Systems: http://necsi.org/.
Gleick, J., (2008) *Chaos: Making a new science.* New York: Penguin Books.

Kellert, S.H., (1994) In the wake of chaos: Unpredictable order in dynamical systems. Chicago: The University of Chicago Press.
http://www.answers.com/topic/catastrophe-theory
http://fan.lib.ru/z/zharow_a/2050buduschee.shtml

Alexandre Monnin

AFNOR http://www.afnor.org/

Bach, E. W. (1989). Informal Lectures on Formal Semantics. New York: State University of New York Press.

Berners-Lee, T. et al. (W3C/MIT, R. Fielding, Day Software, L. Masinter, Adobe Systems) (2005). RFC 3986, http://tools.ietf.org/html/rfc3986

Dennett, D. C. (1990). "The Interpretation of Texts, People and Other Artifacts" in *Philosophy and Phenomenology Research. Vol. 1, Supplement, Fall 1990*

Fielding, R. et al (UC Irvine, J. Gettys, Compaq/W3C, J. Mogul, Compaq, H. Frystyk, W3C/MIT, L. Masinter, Xerox, P. Leach, Microsoft, T. Berners-Lee, W3C/MIT). (1999). RFC 2616, http://www.ietf.org/rfc/rfc2616.txt

FLICKR http://www.flickr.com/groups/api/discuss/72157594497877875/

Floridi, L. (2009). "Web 2.0 vs. the Semantic Web: A Philosophical Assessment" in *Episteme* 6, (1).

Gruber, T. (2005). TagOntology - a way to agree on the semantics of tagging data, http://tomgruber.org/

—. (2007). "Ontology of Folksonomy: A Mash-up of Apples and Oranges" in *Int'l Journal on Semantic Web & Information Systems*, 3(2),

Halpin, H. (2006). "Identity, Reference, and Meaning on the Web," in *Proceedings of the Identity, Reference, and the Web Workshop at the World Wide Web Conference* (WWW 2006), Edinburgh, Scotland, (http://www.ibiblio.org/hhalpin/irw2006/hhalpin.htm)

Halpin, H., and Presutti, V. (2009). "An Ontology of Resources: Solving the Identity Crisis", in *The Semantic Web: Research and Applications*, pp. 521-534.

Hayes, P. (2006). In Defense of Ambiguity, Edinburgh, Scotland: Press

Hottois, G. (1996). Entre symboles et technosciences. Seyssell: Champ Vallon

IRI http://www.iri.centrepompidou.fr/

Limpens, F., and Gandon, F., and BUFFA, M. (2008). "Rapprocher les ontologies et les folksonomies pour la gestion des connaissances partagées: un état de l'art", in *IC2008, 19èmes Journées Francophones d'Ingénierie des Connaissances*, Nancy, France.

Marlow, C., and Naaman, M., and Boyd, D., and Davis, M. (2006). "HT06, Tagging Paper, Taxonomy, Flickr, Academic Article, ToRead" in *Proceedings of Hypertext 2006*, New York: ACM Press.

Monnin, A. (2009a) "Artifactualization: Introducing a new concept", presented at the InterFace 2009: 1st International Symposium for Humanities and Technology, Southampton, United Kingdom, 2009, http://hal-paris1.archives-ouvertes.fr/hal-00404715_v1/ (also available on Scribd with conference proceedings).

—. (2009b), "From Game Neverending to Flickr. Tagging systems as ludic systems and their consequences" in: *Proceedings of the WebSci'09: Society On-Line*, 18-20 March 2009, Athens, Greece (in Press)

Monnin, A., and Felix, E. (2009), "Essai de comparaison des ontologies informatiques et philosophiques: entre être et artefacts", in *XVIèmes Rencontres interdisciplinaires sur les systèmes complexes naturels et artificiels de Rochebrune*, Megève, France.

Newman R. (2005a). Tag Ontology, *holygoat.* http://www.holygoat.co.uk/owl/redwood/0.1/tags/

—. (2005b). Tags, *holygoat.co.uk.* http://www.holygoat.co.uk/blog/entry/2005-03-23-2

Passant, A. (2007). "Using Ontologies to Strengthen Folksonomies and Enrich Information Retrieval in Weblogs", in *Proceedings of the First International Conference on Weblogs and Social Media (ICWSM)*, Boulder, Colorado, USA.

SANDBOX http://sandbox.yahoo.com/VideoTagGame/

Schacter,J. http://en.wikipedia.org/w/index.php?title=Talk:Tag_%28metadata%29&action=edit&section=14

Sinha, R. (2005). "A cognitive analysis of tagging" in *Rashmi's blog*, http://rashmisinha.com/2005/09/27/a-cognitive-analysis-of-tagging/.

Smith, G. (2008). Tagging: People-Powered Metadata for the Social Web. 1st ed., Place: New Riders

Svenonius, E. (2000). The Intellectual Foundation of Information Organization. Place: MIT Press

UPPCOMMING http://www.upcoming.org

Vander Wal, T. (2007). Folksonomy Coinage and Definition. Place: Press

Wittgenstein, L., (1953). *Philosophical investigations*. Place : Press

Zacklad, M. (2007). Classification, thésaurus, ontologies, folksonomies: comparaisons du point de vue de la recherche ouverte d'information (ROI), http://archivesic.ccsd.cnrs.fr/sic_00202440/en/

George Ceausu

Ceauşu, George (2007). "Multy-, Inter- and Trans-Disciplinarity: a Triadic Discourse", in *Journal for Interdisciplinary Research in Religion and Science*, 1 / 2007, (available also on *www.jirrs.org* website)

Chaitin, Gregory J (1964)., "Informatic-theoretic limitations of formal systems", *Journal of Association for Computing Machinery*, 21, pp. 403-424

Constantinescu, P. (1983) *Unitary modelling of system genesis* (Modelarea unitară a genezei sistemelor). Bucureşti: Editura Tehnică, pp. 21-28

—. (1990). *Synergy, information and system genesis. Foundation of synergetics* (Sinergia, informaţia şi geneza sistemelor). *Fundamentele sinergeticii*). Bucureşti: Editura Tehnică, pp. 35 sqq.

Hofstaedter, Douglas R. (1977) *Gödel, Escher, Bach: an eternal golden braid.* New York: Basic Books, pp. 70 sqq.

Hügli, A., and Lübke, P. (2003) (coord.). *Philosophy in the 20th Century.* 2nd volume: *Science theory, Analytic philosophy*, Bucuresti: Editura All

Ioan, P. (2007). (coord.). *Situational paradigm regarding a functional-integrating concept on interpretation (contributions to the inter-disciplinary dialogue).* Paradigma situaţională în perspective unei concepţii funcţional-integratoare asupra interpretării (contribuţii la dialogul interdisciplinar), Exploratory research project ID_413 / 2.10.2007

Lobiuc, I., and Irinescu, T. (2002). *Epistemology and methodology in language science* (Epistemologie şi metodologie în ştiinţele limbajului). Iaşi: Casa Editorială Demiurg

Oprea, D., and Meşniţă, G., and Dumitriu, F. (2005). *Analysis of informational systems* (Analiza sistemelor informaţionale). Iaşi: Editura Universităţii „Al. I. Cuza"

Oprea, D., and Meşniţă, G., and Dumitriu, F. (2006). *Designing of informational systems* (Proiectarea sistemelor informaţionale). Iaşi: Editura Universităţii „Al. I. Cuza"

Neuman, C., and Nicolau, E., and Schor, A. (1983). *Brief history of science development* (Istorie sumară a dezvoltării ştiinţei). Bucureşti: Editura Politică

Rotariu, T., and Iluţ, P. (1997). *Sociological enquiry and the opinion poll* (Ancheta sociologică şi sondajul de opinie). Iaşi: Editura Polirom

Sandu, I. et al. (1994). "Physical effect, physical reality and Substance-Energy-Information model. Elaboration of 15 effects from the inter-disciplinary fund of effects of inventors" (Efectul fizic, realitatea fizică şi modelul Substanţă-Energie-Informaţie. Elaborarea a 15 efecte din fondul interdisciplinar de efecte pentru inventatori), in: *Creative processes, creative models, ways of investigation and application (Procese creative, modele creative, modalităţi de investigare şi aplicaţii)*, Scientific research contract 78B/1994, National Institute for Invention of Iaşi, pp. 150-183

Wirth, N. (1976). *Algorithms + Data Structures = Programs*. Englewood Cliffs: Prentice Hall.

Traian Dinorel Stanciulescu

Allott, R. (1988). "The motor theory of language: origin and function", http //www.percep. demon.co.uk/motorthy.htm, Cortona.

Bârliba, M., Bârliba, C. (1990): *Introducere în epistemologia informaţională* [in Romanian] [An introduction into the informational epistemology]. Bucuresti: Editura Stiintifica, p. 231.

Danesi, M. (1998). *The Body in the Sign: Thomas A. Sebeok and Semiotics.* vol. 1, New York, Otawa, Toronto: Legas Publishing House.

Jackendoff, R. (1987). "On Beyond Zebra: The relation of linguistic and visual information", in: *Cognition.*

Platon (1978). *Cratylos* [in Romanian], in: Opere [Works], vol. 3. Bucuresti: Editura Stiintifica si Enciclopedica.

Popa, C. (1972). *Teoria cunoasterii. Perspectiva semiotico-praxiologica asupra actului cunoasterii* [in Romanian] [The theory of knowledge. A semiotic-praxiologic approach to knowledge act]. Bucuresti: Editura Stiintifica.

Popp, F.A., Li, K. H., Gu, Q., (1989). *Recent Advances in Biophoton Research and its Applications*, World scientific, Singapore, New JerseyLondon, Urban Schwarzenberg, München.

Pribram, K. (1971). *Language of the Brain. Experimental Paradoxes and Principles in Neuropsychology,* 1971, Monteray: Brookes / Cole.

Sebeok, T. A. (1972). *Perspectives in zoosemiotics*. Hague: Mouton.

Stanciulescu, T. D (1996). *Fundamente ontogenetice ale limbajului creativ* [in Romanian] [Ontogenetic fundamentals of the creative language], in: Belous, V., and Stanciulescu, T.D., and Teodorescu, H., and Ungureanu, O. (1996) *Performantica: interferenţe, sinergii, confluenţe* [in Romanian] [Performanctics: interferences, sinergies, confluences]. Iasi: Editura Performantica.

Stanciulescu, T. D., (2002). “On computational knowledge: a semiotic approach”, in: European Conference on Intelligent Systems and Technologies ECIT, Iasi: Editura Performantica.

—. (2003). *Semiotics of light. An integrative approach to human archetypal roots.* Crystal-Concept & Iasi-World Development Organization, Iasi-Geneva.

—. (2005). *The “semiotic graph”: towards an integrative methodology*, in: International Summer School for Semiotic Studies (Abstracts), ISI: Imatra.

Stanciulescu, T. D., and Manu, D. M. (2001). *Metamorfozele luminii: introducere în teoria “laserilor biologici* [in Romanian] Light's metamorphoses. An introduction into the theory of “biological lasers”, Iasi: Editura Performantica.

Teodorescu, H.-N. (2000). “Interrelationship, Communication, Semiotics, and Artificial Consciousness”, in: Kitamura, T. (200). (ed.) What Should be Computed to Understand and Model Brain Functions?, FLSI Book Series, vol. 3, World Scientific.

Voronin, S. “Approaching the iconic theory of language origin”, http: // baserv.uci.kun.nl/~los/Articles/voron.html.

LIST OF CONTRIBUTORS

Alexandru BOBOC
Corresponding member of the Romanian Academy.
Fields of interest: history of philosophy, philosophy of culture and of values, semiotics and philosophy of language.
Researches: Descartes, Leibniz, Kant, Hegel, Kierkegaard, Schopenhauer, Nietzsche, Dilthey, Husserl, Scheler, N. Hartmann a.o)
Papers in: „Kant-Studien"; „Kant-Bibliographie"; „Leibniz-Bibliographie"; „Nietzsche-Studien"; „International Bibliographical Centre" (Cambridge, England); „Revista de filosofie"; „Revue Roumaine de Philosophie" etc.
Member of The Society for Classical Studies in Bucharest, of The Germanists Association in Romania, of the "Kant" Society in Bonn/Mainz, of the "Leibniz" Society in Hannover.
Books (selection): Filosofia contemporană (The Contemporary Philosophy) (1995); *Limbaj şi ontology* (Language and Ontology) (1997); *Cunoaştere şi comprehensiune* (Knowledge and Comprehension) (2001), *Formă şi valoare* (Form and Value)(2005)

Piotr (Peter) BOLTUC, Associate Professor of Philosophy, University of Illinois Springfield and University Professor of E-Learning, Warsaw School of Economics. Editor of the APA Newsletter on Philosophy and Computers; co-editor of International Journal of Machine Consciousness and other journals. Published over 50 articles in the areas of machine consciousness, moral-political philosophy and e-learning, including "Reductionism and Qualia." *Epistemologia;* "The Philosophical Problem in Machine Consciousness," IJMC; "The Engineering Thesis in Machine Consciousness," (*Techne*); "Business Ethics: Sins that Originate from One's Not Knowing Quite How to Sin." *E-Mentor;* "Why Common Sense Morality is Not Collectively Self-Defeating." *Polish Journal of Philosophy;* "Global E-Learning Environment in Philosophy." *Dialogue and Universalism;* "Global Learning Through Collaboration." *Formamente* Presented over 100 lectures including five at the APA and 12 at the World Congresses of Philosophy. Visiting fellow at Princeton and Oxford.

Mihai FLOROIU has graduated in International Law at the Paris I Pantheon-Sorbonne University in 2000 and in 2004 obtained his doctoral degree in law at the University of Craiova with a dissertation on the pluridisciplinary characteristics of the international treaty. Holder of courses on international law at the University of Craiova. Academic expert and deputy project manager of the Romanian Agency for Quality Assurance in Higher Education - ARACIS. Author of three courses on international public and private law and of several studies and articles on various themes in Law, Philosophy and Quality Assurance.

Mircea FRUNZA, born in 1965 (Iasi), is graduate in Mechanical Engineering (Machine Manufacturing and Industrial Management) by the Techinal University "Gheorghe Asachi" Iasi. From 2004, Ph.D. in Industrial Engineering. Has participated in 15 research projects. From 2005 he is working for the Technical University "Gheorghe Asachi" from Iasi and the National Inventics Institute as a Senior Research Engineer. He published 30 scientific papers, has three inventions patented, and the received five gold, tree silver medals and many prices at International Inventions Saloons.

Emilia GULICIUC is Professor of *Philosophy, Global Studies* and *Cultures & Civilization* at *Stefan cel Mare* University of Suceava, Romania.
Pioneer of the online publishing and Internet research in Romania – former director of ROSLIR – Romanian Cultural Studies Internet Review she is member of various professional groups: member of the *UNESCO network of women philosophers;* member of the Romanian Quality Assesment Authority (ARACIS); co-initiator and board member of the *Romanian Society of Philosophy, Engineering and Technoethics.*
Author of 11 books and 60 articles on contemporary and Romanian philosophy, under the theme of the *philosophy immersed in the common human activities.*
Selected publications from the last three years: "The Universalities of Being"; "Modernism, postmodernism, transmodernism and national identity"; "Les universalités de la dimension roumaine de l`être".

Viorel GULICIUC is Professor of *Philosophy, Semiotics* and *Digital Libraries* at the University of Suceava, Romania and associate Researcher of the Indian Institute of Science and Religion of Pune.
Pioneer of the online publishing and Internet research in Romania – ex: director of ROPHIDAS – *Romanian Philosophical Databases Systems* – granted by the Romanian Academy) he is member of various professional

groups: co-initiator and president of the *Romanian Society of Philosophy, Engineering and Technoethics;* board member of the Romanian Semiotic Studies Association (AROSS).
Author of 11 books and 60 articles, under the themes of philosophy and IT, he has also published an extended study in the *semiotics of the philosophical discourse (1998),* for the defence and promotion of the ideas of *non-generic universality* and *un-unitary being.*
Selected papers from the last 3 years: "How Do We Need Universities in a Technological World?"; "The Presupposition of Universality"; "The Wondering Angels of the Fractal Art".

Ionut ISAC
Doctor in Philosophy (Ontology and Metaphysics) at the University "Babes-Bolyai", Department of History and Philosophy (1996);
He is senior researcher Ist class at the Institute of History "George Barit", Department of Philosophy, Romanian Academy, branch of Cluj-Napoca.
Main fields of specialization: Philosophy of Science, Ontology, History of Romanian Philosophy.
Member of the Philosophical Association "Ferdinand Gonseth", St.-Imier, Switzerland (1993); International Foundation for Science and Culture "Stefan Lupascu", Iasi, Romania (1997); Romanian Academy of Scientists (1999); French Philosophical Society of Romania (2000); international commission for nominating candidates to the Inamori Foundation's prize in creative arts & moral sciences, Kyoto, Japan (1996, 2000 and 2004) a.o.
Scientific works: 19 books and about 300 articles, studies, essays published in Romania and abroad; more than 100 papers at congresses, conferences, symposia and colloquia; several research fellowships.

Lorenzo MAGNANI
Philosopher and cognitive scientist, Lorenzo Magnani, is a professor at the University of Pavia, Italy, and the director of its Computational Philosophy Laboratory. He is visiting professor at the Sun Yat-sen University, Canton (Guangzhou), China. He has taught at the Georgia Institute of Technology and at The City University of New York and currently directs international research programs in the EU, USA, and China. His book *Abduction, Reason, and Science* (New York, 2001) has become a well-respected work in the field of human cognition. The recent book *Morality in a Technological World* (Cambridge, 2007) develops a philosophical and cognitive theory of the relationships between ethics and technology in a naturalistic perspective. The last book *Abductive Cognition. The Epistemological and Eco-Cognitive Dimensions of*

Hypothetical reasoning is published by Springer, Berlin/Heidelberg/New York (2009). In 1998, he started the series of International Conferences on Model-Based Reasoning (MBR).

Sorin-Tudor MAXIM, PhD (1998) at "Al. I. Cuza" University. Professor of Philosophy, Ethics, and Political Sciences at *Ştefan cel Mare* University, Suceava and Chancellor of this University.
Main fields of scientific activity: Moral philosophy, Political philosophy Publications: 15 books and 70 papers: *Conştiinţa morală* (The Moral Conscience), *Responsabilitatea morală* (The Moral Responsibility), *Toleranţa. Dreptul la diferenţă* (Tolerance. The Right to Difference), *Mileniul III. Dialogul umanismelor?* (The Third Millenium. The dialogue of Humanisms? (Co-author with Bogdan Popoveniuc); *2001: Umanism şi Educaţie* (2001: Humanism and Education)(Co-author with Viorel Guliciuc); *Secolul XXI. Orizonturi umaniste* (The 21st Century. Humanistic Horizons) (Co-author with Alexandru Ştefan Băişanu), *Violenţa în sport* (Violence in Sport) (co-coordinator, with Dan Ioan Dascălu, Bogdan Popoveniuc and Eusebiu Ionescu).

Dan L. MILICI Doctor of Science: Electrical engineering - Instrumentation measure, specialized in metrology, sensors and microsystems for transducers, interfaces and data acquisition applications of numerical calculations in digital processing of information. Associate Professor in the Faculty of Electrical Engineering and Computer Science from the University "Stefan cel Mare" Suceava Teaching and research activity in the areas: numerical methods and data used, non-electrical quantities measurement, measuring microsystems, virtual instrumentation and monitoring systems and transmission data (Masters) and research in technical education at a distance (e-learning) and automated measurement systems in industrial or social backgrounds.

Mariana R. MILICI Doctor of Science: Electrical engineering - Instrumentation measure, specialized in metrology, sensors and microsystems for transducers, electrical circuits and digital field simulation. Associate Professor in the Faculty of Electrical Engineering and Computer Science from the University "Stefan cel Mare" Suceava, Teaching and research activity in the areas: simulation of electrical circuits, electromagnetic field modelling, signals and systems, electrical circuits.

Virgil MOLDOVAN is a consultant professor at the Technical University from Cluj Napoca Romania (UTCN). Engineer doctor diploma (PhD) at the Czech Technical School (CVUT) in Prague. Main fields of research: technical, industrial aesthethics, designing theory, finit elements and fuzzy calculus, philosophy of technology. Over 120 scientific works, 5 books, 10 didactical works, 59 scientific reports, 11 inventions.

Alexandre MONNIN is a PhD student at Panthéon-Sorbonne University (Paris I), associate researcher to the National Conservatory of Arts and Crafts (CNAM) and former chief librarian and research fellow at the Ecole nationale des chartes. His work in the nascent field of the "Philosophy of the Web" deals with the changes the Internet and its most popular extensions that have recently brought to philosophy. Problems of meaning, reference and identity were all given a new life thanks to the Web and the Semantic Web albeit in a radically new context, full of technological artefacts. In the digital era we live in, these ubiquitous mediators transform everything we know from personal identity to language, meaning or knowledge. Artefacts nowadays both embody such concepts and notions that are integral part of humanity while living a life on their own. For Alexandre Monnin, it is the task of philosophy to understand the consequences of this evolution.

Florin MUNTEANU
Coresponding Member of the Academy of the Scientists from Romania, Scientific Secretary of the Geonomic Sciences Department, Associate professor within the University of Applied Sciences of the Polytechnic University from Bucharest, Senior researcher in the Institute of Geodinamics of Romanian Academy, Scientific Secretary of Unesco Chair in Geodinamic. Founding President of the Center for Complexity Studies – UNESCO center. Initiator of the research programs developed under the general program NEXUS (open projects). *Areas of specialty:* Consultancy and mentoring for training in the human resources for a knowledge based economy (mind building).

Dr **Kuruvilla PANDIKATTU** SJ (1957-) Reader, Department of Philosophy, Jnana-Deepa Vidyapeeth, Pontifical Institute for Philosophy and Religion. He is a multifaceted personality. Holder of three Masters Degrees and two doctorates (in philosophy and theology) from the University of Innsbruck, Austria, he has written extensively on topics of science, religion and philosophy. He has written more than 24 books and 75 academic articles in peer-reviewed journals. He has been involved in the dialogue between science and religion and between religions. His main areas of specialization are: Dialogue as a Way of Life, Hermeneutics and Philosophical Anthropology. Having organized numerous national and international conferences, he has travelled around the world (including US, UK, Germany, Spain, Portugal, Malta, Sri Lanka and Korea) to give lectures and take part in various academic activities. He has been awarded John Templeton Science-Religion Course Award, ICPR Fellowship and his articles have received special prizes. The focus of his academic engagement has been on: Issues in Philosophical Anthropology, Science-Religion Dialogue and Viable Life-Styles. He has also contributed to the role of science and technology towards fostering life.

Boris PLAHTEANU PhD (1973) Professor at Technical University "Gheorghe Asachi" Jassy; Academic activity (didactic and research) at the Technical University "1962-2006, and the National Institute of Inventics: Head of department since 1990, and since 2002 general manager; Doctorship Guide –"Machine-tools" (since 1990).
Fields of interest: Integrated manufacturing systems; Bases of technical creation; Values in engineering; Processes and equipments for nonconventional processing. Activity: project manager in over 65 projects of research, designing, technological development and grants, 28 books, 345 scientific papers published, 75 patents. *Distinctions:* The Order "Steaua Romaniei"(Romania's Star), The Commander Cross of the Order "The Inventive Merit", Brussels, The Great Officer of the Order "Gogu Constantinescu", The Officer Cross of the Order "The Belgo-Spanish Merit.

Bogdan POPOVENIUC PhD in Philosophy at "Al.I.Cuza" University of Iasi with a dissertation on Kantian antinomies of space, time and matter in the epistemology of modern sciences. He also studied Psychology and has a master's degree in "Psychotherapy and social intervention" and in Philosophy. Now he is the Head of Department of Philosophy, Social and Political Sciences at University „Stefan cel Mare" of Suceava, where he teaches courses in Philosophical, Political and Cultural Anthropology, Introduction in Philosophy and Psychology. He is also author of several books including "Ecce Homo!?" and "The Illusions of Reason", and he published several studies and articles on various themes in Philosophy, Epistemology, Anthropology and Psychology.

Cătălina-Daniela RADUCU is a PhD. Lecturer and Researcher at the "Gh. Zane" Economic and Social Research Institute of the Romanian Academy. Her first book, *The Ontological Pluralism and the Idea of Commonsense* was published in 2007. She is also the author of several studies in national and international volumes (ex.: *The Human Being in Contemporary Philosophical Conceptions*, Cambridge Scholar Publishing, 2009) and scientific journals. Her main areas of interest are epistemology, logic, gender and law studies. She is a member of the UNESCO International Network of Women Philosophers and of the Romanian Society of Philosophy.

Colin T. A. SCHMIDT is a specialist in the Epistemology of artificial and natural intelligence as well as the relation between them. As such, he is interested in Human-Computer Interaction, Humanoid Robotics, as well as novelties in Virtual Reality and Communication. In order to further his research agenda he draws on universal notions in the Philosophy of Mind and Language such as Intentionality, reference and categorisation. His approach is often anthropological in nature. He focuses on discourse and logic about our future. Utilitarianism is important in the analyses this author makes; he is concerned with the Ethics of Science. His methodology is always deliberately probatory and terminological.

Traian Dinorel STANCIULESCU (b. 1951), graduated from the Institute of Architecture (Bucharest) and from the Faculty of Philosophy (Iasi), Philosophical Doctor (specialization: logic & semiotics), is a full professor at the Department of Science of Communication at "Al. I. Cuza" University of Iasi, where he teaches and coordinates doctoral papers in the field of semiotics and creatology. He is a senior scientific researcher at the National Inventics Institute Iasi and President of Romanian Association for Semiotic Studies. He has (co)authored 30 books, among them: *Myths of creation: semiotic lectures* (prize the of the Romanian Academy); *Biophotonics fundamentals* (2002); *Therapy by light* (2003); *Semiotics of light* (2003); *Signs of light* (2004); *Semiotics of love* (2007); *El poder de la luz* (2007), *Biophotonic clothing* (2008).

Cristian SUTEANU M. Phil. University of Bucharest, Ph.D. Romanian Academy, Bucharest. Dr. Suteanu is an Associate Professor of Geography and Environmental Science at Saint Mary's University in Halifax, Canada His main research and teaching interests concern informational processes involved in environmental systems and human-environment relations. His research topics include nonlinear systems analysis and modelling, space-time patterns of environmental change and natural hazards, and informational aspects of patterns and pattern change; he is also exploring the implications of these studies for transformations in our cultural meanings and our approaches to information management.

Raymond TURNER
Professor of Logic and Computation, University of Essex 1985- (lecturer 1973-85, dean Graduate School 1999-); Sloan fellow in cognitive sciences University of Mass 1982 (senior research fellow 1986 and 1989), visiting professor University of Rochester NY 1982, visiting fellow Centre for Study of Language and Information Stanford University of California 1984 (consultant in sciences 1982), visiting professor University of Texas Austin 1987.
Editor of *Logic and Computation, Stanford Enclopedia of Philosophy*.
Books: Logics for Artificial Intelligence (1984), Truth and Modality for Knowledge Representation (1990), Constructive Foundations for Functional Languages (1991), Computable Models (2009).

INDEX

Index of Names

General Index